SAAB
VIGGEN

Herbert Lugert
Heimbaustr. 7 a, T. 82462
8900 Augsburg 21

SAAB VIGGEN

Modern Combat Aircraft 21

Robert F. Dorr

IAN ALLAN LTD
LONDON

Contents

First published 1985

ISBN 0 7110 1484 1

All rights reserved. No part of this book may be reproduced or transmitted in any form or by any means, electronic or mechanical, including photo-copying, recording or by any information storage and retrieval system, without permission from the Publisher in writing.

© Robert F. Dorr 1985

Published by Ian Allan Ltd, Shepperton, Surrey; and printed by Allan Printing Ltd at their works at Coombelands in Runnymede, England.

Introduction 5
1 **Guarding the North** 6
2 **Swedish Airplanes for Sweden** 14
3 **Draken** 23
4 **Design, Development and Flight Test** 35
5 **Technical Description** 54
6 **Viggen Variants** 68
7 **Flying the Viggen** 87
8 **The Viggens that Weren't** 97
9 **'It's a Combat Situation Out There . . .'** 103
Specification 111

Introduction

It is beauty and potency. It is designed to succeed in the crucible of air combat. It is Viggen – a plane, a fighter, a triumph by a manufacturer long known for excellence but only now stepping on to the world stage. You will fly in that manufacturer's first major airliner soon, and often, as SAAB's pedigree becomes better acknowledged outside Nordic climes. But Viggen is flown only by special men – fighter pilots who can face heavy odds and prevail. Viggen can fight and win.

I sat in it. The sun poured down on the flight line and the Viggens looked ready to go, straining at the bit, held to earth only by the umbilical of the APU. I braced my feet on the rudder pedals, unusually large and solid. I peered through the clear windscreen and admired the high, all-round pilot's view. My hand closed around the stick. I sat in that Viggen for a long time and admired it with a deep-felt intensity. I wanted to borrow it and try it out.

Clint Eastwood wanted to borrow one for his film *Firefox*, to use Viggen to represent a ficticious modern fighter. No airframe could be made available, and so Hollywood had to settle for a special-effects replica instead. It seems unfortunate: the spectacular looks and remarkable performance of Viggen might have been a box-office draw.

Not that Viggen is perfect. It's short-legged, most believe. Early in their career, AJ37 Viggens were grounded with wing spar problems. In February 1984 they were grounded again because cartridges on some rocket ejection seats had been installed upside-down, so all had to be X-rayed. Viggen has had other teething problems. But none deters from the simple fact that Viggen is among the most remarkable aircraft of its era.

Any Viggen pilot, and most Swedish aviation enthusiasts, could have written this book better. They shall. The full story of Viggen will not fit in a volume this size and will not reach its conclusion in this century. Because the story isn't over yet, Viggen's impact upon our world and our time will be measured by others in some future day. This is a modest beginning, a story of an airplane for those who love airplanes and for those who fly and fight.

Photographs for this book were provided by Saab-Scania, Volvo Flygmotor, and by the Swedish Air Force. Hans G. Andersson, a unique mixture at SAAB, and John Charleville of Stockholm's air staff were generous in providing pictorial content. The text is based on published materials and material provided by dozens of enthusiasts and historians who gave of their time and effort. Any errors are the responsibility of the author.

I would like to acknowledge my special indebtedness to Michael A. France, friend, mentor, who knows more about fast jet combat aircraft than almost anyone.

I am also indebted to Bengt Andersson, Robert J. Archer, Paul Auerswald, Lennart Berns, Roy Braybrook, Robert L. Burns, John Charleville, Robert P. Dorr, Lawrence G. Dorr 3rd, Dennis Finnerty, James W. Freidhoff, Bill Gunston, Alan W. Hall, Martin Judge, M. J. Kasiuba, Hans Percy, Bryan Philpott, Chris Pocock, Jerry Scutts, Norman Taylor and Lars Thomasson.

The views expressed in this book are my own and do not necessarily reflect those of the Department of State or of the United States Air Force.

Robert F. Dorr
London

1 Guarding the North

A throaty growl. Then a rumbling sound, which rises in crescendo. Finally, a roar. The sound of jet engines on after-burner shatters the air and shakes the land.

Accelerating from zero to over 150mph (241.2km/hr) in seconds, two lizard-hued Viggen fighters rotate, break ground, and hurtle skyward at 45°, long tongues of flame stabbing behind them. The angle itself defies the mind, few fighter aircraft being able to display such an arrogance towards gravity. It is not enough to say that a pair of Viggens, hurtling aloft, is an impressive sight. On its way up, the Viggen climbs more abruptly than anything of its generation – more than a Jaguar, a Phantom, a MiG-23. No modern combat aircraft so captures the attention of an on-looker during the critical moments after take-off.

No other fighter is as loud. None uses as little runway. The specification for the Viggen calls for operation from narrow strips of 500m (1,640.4ft) length and the aircraft is routinely, comfortably airborne well before the 600m (1,968.5ft) marker. A SAAB advertisement, prepared in 1971 when foreign sales of the Viggen seemed likely, dramatises the

Above:
Standing guard over the coastal approaches to Swedish airspace, four AJ37 Viggens from F15 wing at Soderhamn fly formation in 1977. The aircraft bearing the tail number 43 has recently been transferred from K6 wing at Karlsborg and does not yet have its new unit's number painted on the nose. *Saab-Scania*

rapidity with which the Viggen soars to fighting altitude. The aircraft 'takes less than 100 seconds to climb from take-off to an altitude of 10 kilometers (32,808ft). What's more, Viggen goes supersonic in only 60 seconds from take-off.' This compares favourably with fighters designed nearly two decades later, such as the Northrop F-20A Tigershark, which has a more favourable thrust-to-weight ratio but can only barely exceed the Viggen's impressive climb rate.

In moments, the two Viggens are at fighting altitude over thinly-populated Swedish forestland, climbing out toward the east coast and the contested Baltic Sea.

In 1970, fewer than 25 intercept missions were

flown; in 1981, the figure leaped to 200, with 49 actual airspace violations reported, those by NATO aircraft slightly exceeding those of the Warsaw Pact.

If this is a 'hot' mission, not yet determined to be not the 'real thing', an advanced warplane designed and built at home by a neutral nation could be heading into battle with a potent arsenal – Falcon, Sidewinder and Skyflash missiles, and the best internally-mounted cannon of any fighter in the world. And in the short-notice, high-speed encounters typical over the Baltic, the man flying the Viggen may not know until making visual contact whether this *is* the real thing. It would be an exaggeration to suggest pent-up tension inside the cockpit, but the situation is serious, and the aircraft itself was built for a serious purpose . . .

System 37, as the family of Viggen fighters is called, rules supreme in Nordic skies. A graceful and potent aircraft denied real beauty only because its fuselage seems too short for its delta-canard wings, *Flygplan* 37 is a national symbol of Stockholm's will – the final resort on the cutting edge of Swedish policy.

The Viggen is advanced because the nation which

Below:
The natural-metal finish is no longer usual for the Viggen, but the remote airfield shelter and armament typify *Flygvapen's* **role in defending Swedish air space. In this 1973 photo an AJ37 Viggen taxies out with Rb.24 (AIM-9 Sidewinder) air-to-air missiles under its fuselage and Rb.28 (AIM-4 Falcon) air-to-air missiles under the outer wings.** *Saab-Scania*

Bottom:
Few aircraft are as impressive, taking off or landing, as the Viggen. Few can get into, or out of, smaller spaces more easily. This flight-test example of the AJ37 Viggen carries a collision-avoidance pod beneath its right fuselage. *Robert Lofberg*

built it, comprising a mere eight million people, is itself at the fore of science and technology. The Viggen is designed at home (engine and avionics aside) because Stockholm wishes never again to depend upon foreign suppliers: such dependence in the 1930s followed by the 'shock' of a US embargo in 1940 left Sweden without aircraft it had already purchased and forced the country to build its own. The Viggen wears the triple-crown insignia of a nation which belongs to no alliance and asserts non-alignment, but is more heavily armed than NATO Denmark or Norway. The Viggen belongs to a neutral nation whose people have a healthy understanding of the Russians.

Right:
Viggen pilots are very good – and very young. Beginning in 1985 men with no previous fighter experience will step directly into Viggen cockpits. These Viggen fliers of F15 wing at Soderhamn wear typical flying garb, which is very similar to that worn by Britain's Royal Air Force pilots.
Swedish Air Force

Below:
External power is normally used to start the Viggen, even when on 'alert' status. Here, power hose and umbilicals have been attached to JA37 Viggen interceptor of F13 wing at Norrkoping, seen in rare Swedish sunshine.
Swedish Air Force

To some, the airplane seems like the people who build it – attractive, functional, with a no-nonsense dedication to practical success and with, it must be added, a certain lack of flair. Though it looks beautiful to an engineer, the Viggen is not known to have inspired poets. If some airplanes evoke love and passion, if some take on human qualities, the Viggen makes you think how well men build machines. You are not likely to become romantic toward this aircraft, but you're certain to bestow it ungrudging respect.

'Respect,' says Lennart Berns, a Swedish Air Force major. 'That's what we must have. And *do* have. That's why we defend our skies, as any sovereign nation would.'

The Swedish Air Force

Sweden's armed forces are steered from a Supreme Command in Stockholm which reports to the Prime Minister through the Defence Minister. In 1984, the all-important Defence portfolio was held by Anders Thunborg, a respected figure attuned to the importance of airpower. The *Svensk Flygvapen*, or Swedish Air Force (from which the modifier *Kungliga*, or Royal, was dropped in 1974 although the nation remains a monarchy) keeps its headquarters at the Air Staff, or *Flygstaben*, in Stockholm just off the divided, tree-lined boulevard Valhallavagen. General Sven Olson, at age 55 fully qualified in the Viggen, is chief of an air force which has dwindled in size from 50 fighter squadrons at its heady peak in 1950 to possibly as few as 12 after 1986. Olson's Air Staff is a lean and dedicated body of professional officer/technocrats whose challenge is to assure an air arm of top quality now that rising costs have made unavoidable a drawdown in quantity. To guard the North, their means is the largest budget of any of the Swedish services. Their instrument is the Viggen. Even the arrival in the late 1980s of the supremely advanced JAS39 Gripen will not, at least at first, denude the primacy of the System 37 Viggen in defending Nordic skies.

From red-brick office buildings in Stockholm, the profession of aerial arms branches out to six regional military commands, or *militaromrade*, each with a geographical responsibility: North Norrland, South Norrland, East, South, West and Bergslagen. Scattered throughout the country are 15 wings, or *flotilljer*, each headquartered at a major airbase.

Far more than in other services – Britain's Royal Air Force, for example – the identity of base and wing is intertwined, one man commanding both, one command structure administering both, and many wings trace their history back to the beginning of the Swedish Air Force as an independent service in 1926. Each *flottilj* operates two to four *divisioner* (squadrons) of 15 to 18 aircraft and each is divided into three sections: flying, services, and maintenance. The air force numbers about 19,000 people, combining a dedicated professional corps with 'citizen-soldier'

Above:
Once camouflaged to protect them while on the ground, Viggens now wear a greyish camouflage scheme designed to make them difficult to see in flight. Here a JA37 interceptor of F13 wing at Norrkoping makes an abrupt take-off in the new grey paint scheme. *Swedish Air Force*

conscripts. Because it serves in several key roles – attack, sea surveillance, reconnaissance and air defence – the Viggen is a familiar sight throughout the service.

A major innovation of the Swedish forces is the short take-off/landing (STOL) capability which permits the Viggen to operate from dispersed locations, including roadways. Both the Draken and Viggen were planned with STOL in mind, the Viggen being able to stop in 500m (1,640.4ft) using its reverse thrusters. Short landing is accomplished by the use of automatic throttle, head-up display (HUD) for precision glidepath adjustment and, of course, thrust reversal – a carrier-style landing without flare being routine. In a crisis, the aircraft would be deployed to pre-planned road locations. Several road bases have hardened fuel and storage facilities, and elsewhere, widened and reinforced highway pavement provides a 'runway' even where other facilities are lacking. Some aircraft in each *division* are on 'alert' for rapid deployment to these less vulnerable locations. Gray camouflage for air-to-air visibility impairment is gradually replacing the mottled green-brown used to camouflage the aircraft on the ground. But when Viggens still wearing the four-colour, green-brown scheme are shielded beneath foliage, and when readiness is being tested, the casual passer-by can be astounded by the sight of a Viggen taxying out of the trees! STOL and its close relative, V/STOL, are ignored by other nations at their peril, especially in a time when fixed aerodromes are increasingly vulnerable to nuclear weapons as well as modern non-nuclear ordnance.

Right:
Basking in rare sunlight at Norrkoping, a row of almost factory-new JA37 Viggen interceptors of F13 wing is depicted in about 1982. The JA37 interceptor variant is distinguished from other Viggen models by the sweptback tip atop its vertical tail. *Swedish Air Force*

Below:
Dispersal to rough roadway strips is a key capability of the Viggen, essential to wartime planning. Here, in 1973, a JA37 ground-attack Viggen of F7 wing at Satenas taxies on a roadway. *Saab-Scania*

Below right:
A JA37 interceptor Viggen demonstrates its rakish appearance in a low altitude, high-speed pass. The radar homing and warning system (RHAWS) which tells a pilot he is being stalked by enemy radar, is located in protuberances at the 'break' on the outer wing.
Swedish Air Force

To guard its Nordic approaches, Stockholm as far back as 1950 saw the need for an integrated surveillance and warning network. While the Viggen was taking form, STRIL 60 (*Stridsledning och Luftbevakning* 60) was developed as the nation's radar defence network, providing surveillance and early warning through a system of radars operating at various ranges and altitudes, down to tree-top level. Long-range, three-dimensional radars for surveillance and ground-controlled intercept (GCI) have since 1981 supported the interceptor version of the Viggen, the JA37, in its defensive role. The STRIL 60 system absorbs, evaluates and disseminates data on potential targets not only to fighter-interceptor squadrons but also – in a highly integrated command network – to missile, anti-aircraft, and civil defence units. Ground-based GCI operators can direct Viggen interceptors racing up to meet unidentified aircraft.

The Viggen Pilot

In the end, of course, it is the man inside the machine that matters. The men who sit at Viggen controls are, on average, younger than fighter pilots in most Western air arms. In fact, a conscious effort is under way to lower the age of pilots; in 1984 the Swedish Air Force removed a number of majors and lieutenant colonels from flying positions many had held for 10 years or more. Beginning in 1985, men lacking

Left:
The setting is the historic Lacko Castle on the shores of Lake Vanern, where two AJ37 Viggens of F7 wing at Satenas hold formation at low level in 1973. F7 was the first unit to equip with the Viggen and continues to operate the ground-attack AJ37 aircraft. *Saab-Scania*

previous fighter experience will fly the JA37 interceptor-version Viggen at the outset of their careers, ending a tradition where 1,000 logged hours in the J35 Draken were a pre-requisite – and further cutting the age of the average Viggen pilot. The Swedish fighter pilot also flies fewer hours per year than other Western fighter 'jocks'. A career officer pilot is expected to log a total of some 500 cockpit hours before he is considered fully qualified for tactical operations and normally flies about 150 hours annually, a remarkably low figure which is augmented by experience in one of the world's best ground simulators.

Youth? Inexperience? It is equally true that with the passing of a handful of senior veterans from Congo operations in the 1960s, no Swedish fighter pilot has had combat experience. No Viggen has ever fired a cannon shell or a missile in anger.

But anyone intruding into Sweden's claimed 12-mile territorial limit might be advised that youth, low logbook hours and lack of combat experience do not add up to the sum they imply. That would be as foolish as under-rating the Viggen force because the aircraft is maintained by conscripts which, as noted earlier, it is. With career non-commissioned officers (NCOs) overseeing them, the men on the ground reflect a national service system in which every able-bodied man over age 18 must serve a limited period. Initial training takes 18 months and is followed by regular refreshers every three months with further service up to the age of 45. Sweden could call up 900,000 men in a national emergency and in the meanwhile a cadre of crack officers and NCOs make certain the Viggen force is ready to fly and fight. The conclusion is the opposite of what nearly all of the evidence suggests: fewer fighter forces are as lean, as capable, or as effective as the Viggen force. Few men are held in as much respect as the men who sit in Viggen cockpits.

Consider the near-awe with which the Viggen fighter pilot is regarded by his NATO couterpart. In October 1982 four General Dynamics F-16 Fighting Falcons of the Netherlands Air Force visited Sweden to conduct joint demonstrations at the Viggen base at Angelholm, home of F10 wing. There was much good-natured 'zapping' – quick-handed men with paint brushes applying their own unit insignia to the other guys' flying machines. Some mock dogfighting and some raising of glasses over the O Club bar was inevitable. A Dutch pilot, more mature and experienced, flying a far newer machine, described a younger Swede in the older Viggen with the strongest statement of respect one can utter. 'These guys belong among the very finest fighter pilots in the world. If I were in trouble, I would want that Viggen flier on my wing.'

Respect among fighter pilots transcends boundaries. To balance their infrequent visits by NATO neighbours, Swedish fighter wings on occasion swap courtesy visits with their Soviet counterparts. The last such visit occurred in August 1981 when seven MiG-23s – part of a demonstration flight, devoid of ordnance and stripped of key avionics – accompanied a Tu-154 transport on a goodwill visit to F17 wing at Ronneby. Viggens and MiG-23s did not fly in formation during this get-together and 'zapping', back-slapping and bar-room stories were not major components of the event. Indeed, an air of formality reigned as Soviet fighters basked in rare Swedish sunshine. This time the visitors, unlike the Dutch F-16s, were from an earlier vintage of fighter technology and it is thought that any future exchange in the other direction will be flown again by Drakens, – not the more advanced Viggen. But Viggen and MiG-23 pilots *did* have rap sessions, did compare notes, and shared the flying tales which always come forth when men in cockpits meet and talk – especially when a MiG-23 landed 'short' on the Ronneby strip and was saved from accident only by skilful ground manoeuvring. A Swedish officer says, 'We respect them. They respect us. I never heard a Russian suggest that we were too young or too inexperienced.' It seemed apparent that the Viggen, not conceived from the outset as a close-range dogfighter, could out-turn and outmanoeuver the MiG-23, as well as being able to run away from it on afterburning.

Says Captain Melvin Pobre, a US Air Force tactics analyst, 'I would rate Viggen pilots on a par with our (American) best. They have what it takes.'

The Machine
'I found it remarkable to be standing there amid these big, chunky fighters, the shadows cast by their unique canard double wings etched harshly upon the concrete hardstand by the brilliance of the Scandinavian sun.' So wrote British journalist Roger Lindsay, suggesting that Viggen may inspire poetic language after all. Chunky. If Viggen is not wholly graceful, it may deserve better. But visitor Lindsay's impression speaks of a fighting machine which can, and does, guard the North. Viggen's flaws – there are some – do not detract from its impressiveness or its readiness. The aircraft *is* ready to do what men and machine inevitably must do: fly and fight.

There was a time when Sweden did not enjoy an impressive, home-built aircraft fleet or a state-of-the-art air defence system. In 1940, 7,000 miles from Stockholm, in the White House, a few strokes of a President's pen began the chain of events which led to today's fiercely respected Viggen force.

2 Swedish Airplanes for Sweden

If there was a moment when today's Viggen became inevitable, it may have been when President Franklin D. Roosevelt, with a few strokes of his pen on 10 October 1940, embargoed all shipments of combat aircraft destined for overseas customers.

In Stockholm the reaction was traumatic. Europe was deeply embroiled in war, even if the United States was not. Swedish planners felt that neutrality could be protected only through a military build-up which would make aggression expensive. The cost of unpreparedness had been pointedly brought home six months earlier on 9 April 1940 when General Englebrecht's 163rd Wehrmacht Infantry Division took over Oslo airfield, launching the occupation of adjacent Norway. Roosevelt's embargo undercut a plan by the *Kungliga Svensk Flygvapen*, the Royal Swedish Air Force as it was then called, to significantly increase its fighter and bomber strength. The order prevented Sweden from receiving 60 Seversky EP-1 and 144 Vultee 48C Vanguard fighters, as well as 50 Seversky 2-PA dive bombers, most of which were eventually to be pressed into American service under the designations P-35, P-66 and AT-12.

This caused a 'panic situation' in Stockholm, as one officer recalls it. A vast but sparsely populated nation, armed but neutral, in an isolated corner of the North, was stunned into the realisation that it couldn't rely for help on anyone outside its own borders. 'It was the embargo that decided us', says retired general Nils Soderberg. 'We knew from that moment on that we must design, develop and put into service our own aircraft, from our own factories, so as never to be dependent again . . . '

Though it was the 1940 embargo which led to an insistence on 'Swedish airplanes for Sweden', *Flygvapnet*'s generals might have learned the same lesson from earlier events, including a 1925 recommendation by experts, a 1939 entreaty by SAAB, and the dark comedy of 1938–39 when the generals encountered only grief in their search for a long-range reconnaissance aircraft.

In 1925, the year before Sweden's air force became an independent service, an ad hoc policy group from the Army and Navy had recommended strongly that Sweden manufacture its own aircraft engines. On 18 September 1939 SAAB had written to the

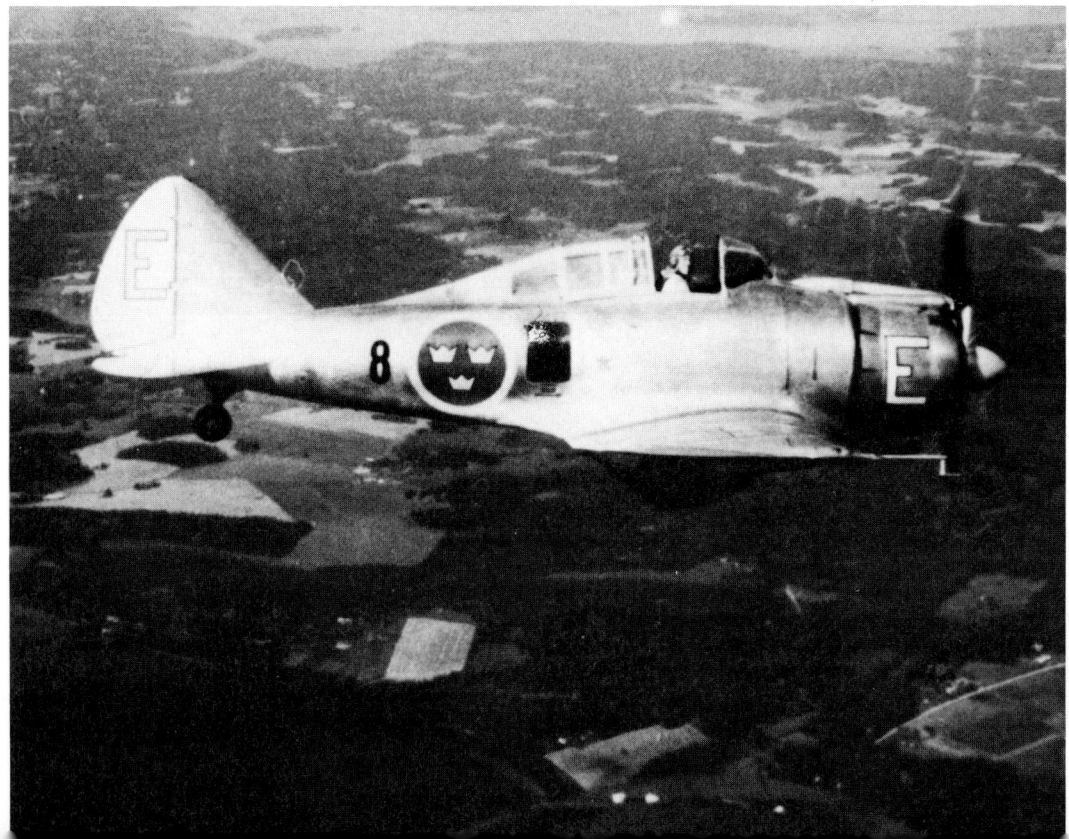

governmental Industrial Commission in Stockholm with a self-serving yet prophetic warning against depending on foreign suppliers of military aircraft.

In 1938 Stockholm attempted to purchase the Blenheim IV bomber for the long-range reconnaissance role only to have its order rejected because of the RAF's own expansion programme. The Swedes then ordered 18 Brequet 694 airplanes as substitutes, but these were confiscated by the French government before they could be delivered. In consternation Sweden ordered 18 Dornier Do215s as a third choice, only to have this order cancelled by the Germans. In their fourth attempt the Swedes ordered 18 Fokker G1s from Holland but that country was overrun before the order could be filled. The long-range reconnaissance portfolio was to remain unfilled until the indigenous SAAB B18 could be used for this role – the same mission assigned in the 1980s to the SF37 Viggen.

First Steps

Svensk Aeroplan Aktiebolag, meaning Swedish airplane company or SAAB for short, had been formed in 1937 to build foreign aircraft under licence. With 'the rug pulled out from under us', as Soderberg describes the choking-off of foreign purchases, a two-tracked approach was taken to free Sweden of its reliance on foreign airplanes. First, under Soderberg's command, a government production team turned out the local J22 fighter, best remembered for its mostly wood structure and its resemblance to the Focke-Wulf FW190. Second, and more important, SAAB began developing a series of Swedish-designed combat aircraft which, in years ahead, would become legend.

'We were never fully independent', admits Hans Westerberg, a fighter pilot in the 1940s. 'We often used American engines and equipment from abroad.

Below left:
Sweden's last major, foreign-built fighter before the 1940 embargo was the Seversky EP-1, given the *Flygvapen* designation J9. This previously unpublished shot of a J9 from F8 wing at Barkaby shows the reception given to Allied bomber crews when they nursed battle-damaged Fortresses and Liberators to internment in Sweden during World War 2. One hundred and twenty of these fighters were ordered by Sweden. Sixty were delivered and, after the embargo, 60 more were diverted to the US Army Air Corps as the P-35. The diversion was so hasty that P-35s were still painted in Swedish markings when, in US service, they were destroyed on the ground during the first Japanese air strikes on Clark Field, Philippines on 8 December 1941. *Robert F. Dorr Collection*

Below:
The result of the 1940 US embargo. Sweden had planned a dive-bomber force of 52 two-seat Severaky 2-PA Model 204 Guardsman aircraft, given the Swedish Air Force designation B6. Only two ever reached Sweden. After the 1940 embargo 50 others were diverted to the US Army Air Corps where they served a lacklustre career as the AT-12 advanced trainer. The sole surviving machine is this AT-12 belonging to the Planes of Fame collection at Chino, California. *William Johnson*

Top:
The J22 fighter of World War 2, remembered by many for its resemblance to the Focke-Wulf FW190, was the first important fighter designed and built in Sweden without outside help. It is also the only important Swedish aircraft type not built by SAAB, having been constructed at the government's FF Verstader Ulvsunda.
Swedish Air Force

Above:
The pusher SAAB J21 fighter began Sweden's tradition of building fighters of indigenous design, which culminated in the Viggen. Each of the native-built aircraft was innovative, and each pioneered new concepts. This J21A stationed with F9 wing at Save on the Swedish west coast would later be supplanted in service by the J21R jet-powered version. *Saab-Scania*

Above:
Based upon the propeller-driven J21, this J21R jet fighter typifies Sweden's determination to design and build its own aircraft. Although this jet variant of the J21 – the only fighter in the world built in both prop and jet versions – was less than a total succes, it paved the way for future generations of native-designed Swedish fighters, including the Viggen. *Robert F. Dorr Collection*

But we were determined to avoid, if possible, being one hundred per cent at the mercy of other governments . . .'

In 1941, when a licence was obtained to manufacture the Daimler-Benz 401 engine, SAAB was ordered to go ahead with a twin-boom, pusher fighter design which became the J21. ('*Jakt*' is the term for fighter, though literally it means hunter.) With the J21 pusher, SAAB began a tradition of manufacturing outstanding Swedish-designed fighters, a tradition which has continued through the Drakens and Viggens of today. All would be of unconventional design. All would incorporate novel features. In each case the company showed an admirable determination in concentrating its efforts to solve the problems and to produce – out of one single project – a first-class fighter as advanced as any in the world.

Some 298 of the prop-pusher fighters were built. Swedish fighters became a familiar sight to hundreds of Allied bomber crewmen who were intercepted and escorted to a safe landing (and internment) after battle damage forced them to turn toward Sweden. Native know-how, good pilots, luck and pragmatic politics got the *Flygvapen* through World War 2 and left the Swedish flying service with a highly experienced corps of fighter pilots even though few, if any, had ever thumbed a trigger in anger. After VE Day Sweden was one of the few countries *not* to disarm.

Determined to maintain strong armed forces to back up its non-aligned political status – and aware that its Soviet neighbour was on the short list of nations *not* hurrying to disarm – Sweden went into the jet age.

SAAB's Jet Fighters
The basic twin-boom design of the J21 was a favourable starting point for a jet venture and both SAAB and the Air Force were anxious to acquire experience in developing jet aircraft. So when a suitable engine, the de Havilland Goblin II, became available in 1946, the jet-propelled J21R was completed, making its first flight on 10 March 1947. It is perhaps worth mentioning that this date was nearly four years after the first flight of the Goblin-powered Lockheed XP-80, a far more advanced design. Sixty J21Rs were manufactured, all assigned to F10 wing at Angelholm. Some were later tested in the ground-attack role as the A21R.

Top:
Contemporary with the F-86 Sabre and MiG-15, the SAAB J29 Tunnan, or Barrel, impressed the world with Swedish technology. Seen in winter snow, aircraft 'Blue I' of F3 wing at Malmslatt, serial number 29484, was built as a J29B, delivered on 22 February 1954, and brought up to J29F standard before it crashed on 21 October 1966.
Saab-Scania

Above:
After the J29 Tunnan entered service, the Swedish Air Force dropped the practice of assigning a letter to each individual aircraft and instead assigned a number to each machine within a wing. Aircraft Nos 10 and 24 of F3 wing at Malmslatt are serial numbers 29490 and 29606 respectively. *Saab-Scania*

Right:
Designed by Lars Brising, who would be SAAB's Vice President for Technology early in the Viggen programme, the J32 Lansen or Lance was an underpowered but successful local design. Aircraft No 33 of F7 wing at Satenas is an A-32A ground attack machine serial number 32085, now retired as a gate guardian at Satenas.
Robert F. Dorr

The combat tactics developed for the J21R were crude and rudimentary, compared with Viggen tactics to follow. They were based on Messerschmitt Me262 experience during World War 2 – that is, hit-and-run and head-on attacks followed by a pursuit curve: the J21R lacked the range and endurance to stay and fight. It had other faults too. Heavy on the controls, slow in roll, with a nose-down tendency at speeds near the critical Mach, the J21R also climbed too slowly to satisfy fighter pilots. To many it might be remembered solely as a curiosity – but the J21R *was* the first jet-powered Swedish airplane for Sweden.

As early as 1945 Swedish Air Force commander Bengt Nordenskjold had decided that all of the force's fighters should be jet-powered. On 1 September 1948 the SAAB J29 Tunnan (Barrel) fighter took to the air for the first time, a swept-wing contemporary of the MiG-15 and F-86 Sabre. Built in fighter, attack and reconnaissance versions, the J29 set an absolute world air speed record on 6 May 1954 with Captain Anders Westerlund at the controls, reaching 607mph (977km/hr), or 13mph (27km/hr) faster than the previous record set by an F-86E. It was used in combat in support of UN operations in the Congo in 1960–64; 661 aircraft of the J29 type were built, more than any other Swedish aircraft.

Few flying machines have been better-loved than the J29, despite its ungainly barrel shape. With its RM2 turbojet engine, a licence-built version of the de Havilland Ghost, the J29 combined 45° swept wings, all movable tailplane, automatic leading edge slots, full span ailerons/flaps, and a newly-designed SAAB ejection seat. RAF Sq Lr Bob Moore, one of the few foreigners to become extensively involved in a Swedish fighter project, carried out 94 test flights on early J29 machines at Linkoping and says that, although the plane may have seemed an 'ugly duckling', it was also 'a pilot's aircraft.' Westerlund, the speed record holder, says that the J29 'was certainly one of the finest fighters of its day, and one which should have been more appreciated.' Armed with four 20mm cannons in the nose and capable of carrying a variety of ordnance stores unders its wings, the J29 was a potent warplane once an early tendency toward buffeting was resolved by a minor change in the shape of the vertical tailplane. No fewer than five J29s have found their way into museums outside Sweden, one being J29F serial number 29640, which belonged to the historical collection at Southend-by-the-Sea until being sold, recently, to a new owner who hopes to refurbish it and place it on display.

Enter the Lansen

In 1956 first deliveries began of the SAAB A32 Lansen attack aircraft, which soon appeared in J32B fighter and S32C reconnaissance variants. Appearing at a time when the Swedish Air Force was at its maximum strength (28,000 men and 11 wings with 50 squadrons), the Lansen (Lance) was among the most aesthetically appealing of fighters. Four hundred and fifty-two were built, some of which remain in service today as target tugs.

By now there existed a well-established and highly-motivated fighter team at Linkoping, the city in central Sweden whose name is almost synonymous with SAAB. These men had taken their country a long way from the 1940 embargo. They were mostly anonymous, these engineers and scientists of SAAB; many were young men and they had greater achievements ahead.

Volumes could be written about the efforts of the SAAB group, which dominated the nation's aviation scene and which, in 1965, employed 14,245 people. The firm has produced all of Sweden's major fighters, but it rates credit for other locally-designed aircraft as well. The SAAB 91 Safir, known as the Sk.50 in *Flygvapen* service, is a highly successful four-seat cabin monoplane used as trainer, transport, and all-around 'hack'. Looking as if it might owe some legacy to Mooney and Beechcraft private types, both of which it actually pre-dates, the Safir, powered by a 180hp (134.2kW) Lycoming O-360-A1A, is still used as a basic trainer. Despite its longevity and utility, the Safir is 'expensive' in the sense that a foreign design, already available in large numbers, would have cost far less to purchase and fly.

The same is true of another local design, the SAAB 105 trainer, known militarily as the Sk.60, a supremely pragmatic machine but one so lacking in inspiration that it has never earned a popular name. Squat, straight-winged, a multi-purpose twin-jet trainer also useful in the light, ground-attack role, the Sk.60 has been built in two and four-seat versions. It has bred a generation of fast-jet pilots, and like the J29 fighter, it has been exported to Austria. It is a 'fun' airplane to fly. Above all, perhaps even more than Viggen, the Sk.60 is a uniquely native product. To be sure, outside thought went into its two Turbomeca Aubisque ducted-fan jet engines rated at 1,640lb (743kg) static thrust. With a wingspan of 31ft 2in (9.5m), length 34ft 5in (10.5m), height 8ft 10in (2.7m), the Sk.60 has an empty weight of 5,534lb (2,510kg) and a maximum take-off weight of 8,930lb (4,050kg). Maximum speed is 478mph (720km/hr) at 20,000ft (6,000m). What is most remarkable about the Sk.60 is that it is a least remarkable airplane. 'They could have purchased almost anything for this job – T-33, Jet Provost, you name it', says an observer. 'They built the Sk.60 from scratch out of sheer determination and pride.'

That Linkoping design team has been kept busy. There are those who say that the Viggen has roots planted even earlier than the roots of the Draken which preceded it. No later than 1 May 1952 SAAB's foremost designer of aircraft, Erik Bratt, began the first of the studies that would lead to System 37, as the family of Viggen aircraft is called. In the middle to late 1950s, SAAB's fighter men argued incessantly over whether 'scientific' knowledge should determine the shape of new fighter aircraft, or whether designers should begin with mission requirements and allow these to dictate the technology selected. Age Roed, one of the design team, felt that the work devoted to a new design had to be, of necessity, as much an art as a science. Bengt Svensson, another SAAB engineer, says that members of the Linkoping team were encouraged to stretch their imaginations, even when the innovative Draken had not yet flown and the more revolutionary Viggen lay in the future. 'All sorts of ideas were put in ink', says Svensson. 'Not just double deltas. Swept-forward wings, flying wing designs, you name it, almost every possible revolutionary idea reached the drawing board stage at one time or another . . .'

The spirit of innovation might be seen, in a sense, as a kind of compensation. Aware that they would be limited in access to outside technology, Swedish designers were compensated with the freedom to explore new ideas.

Still, the notion of 'Swedish airplanes for Sweden' – more than four decades after the US embargo created the phrase – remains a fiction. No nation, at least none but the world's two great superpowers, can produce an advanced, front-line combat aircraft which is 100%

Below:
Test models of the two great fighter aircraft of an era, the J32 Lansen (left) and J35 Draken (right) at SAAB's Linkoping factory, together with some of the test pilots who have flown these machines. Solid experience in carrying out flight test programmes was to be a major boon to SAAB when the Viggen programme began.
Saab-Scania

Above right:
Another design which typifies home-grown efficiency is the SAAB 105, military designation Sk.60. This unique design comes in two and four-seat, trainer and ground-attack versions, and is an important part of the Swedish Air Force today.
Robert Lofberg

Right:
An Sk.60 trainer banks over downtown Stockholm, the capital's city hall just visible above its tail. *Robert Lofberg*

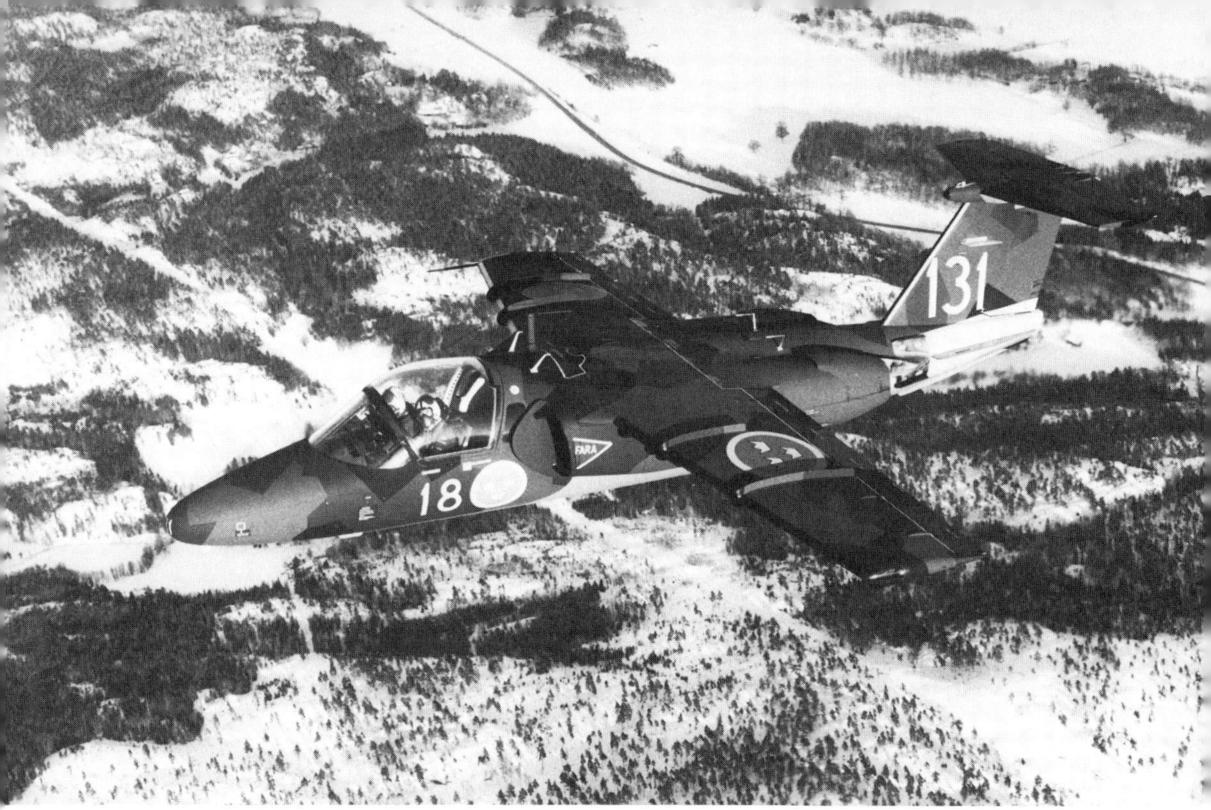
131
18
FARA

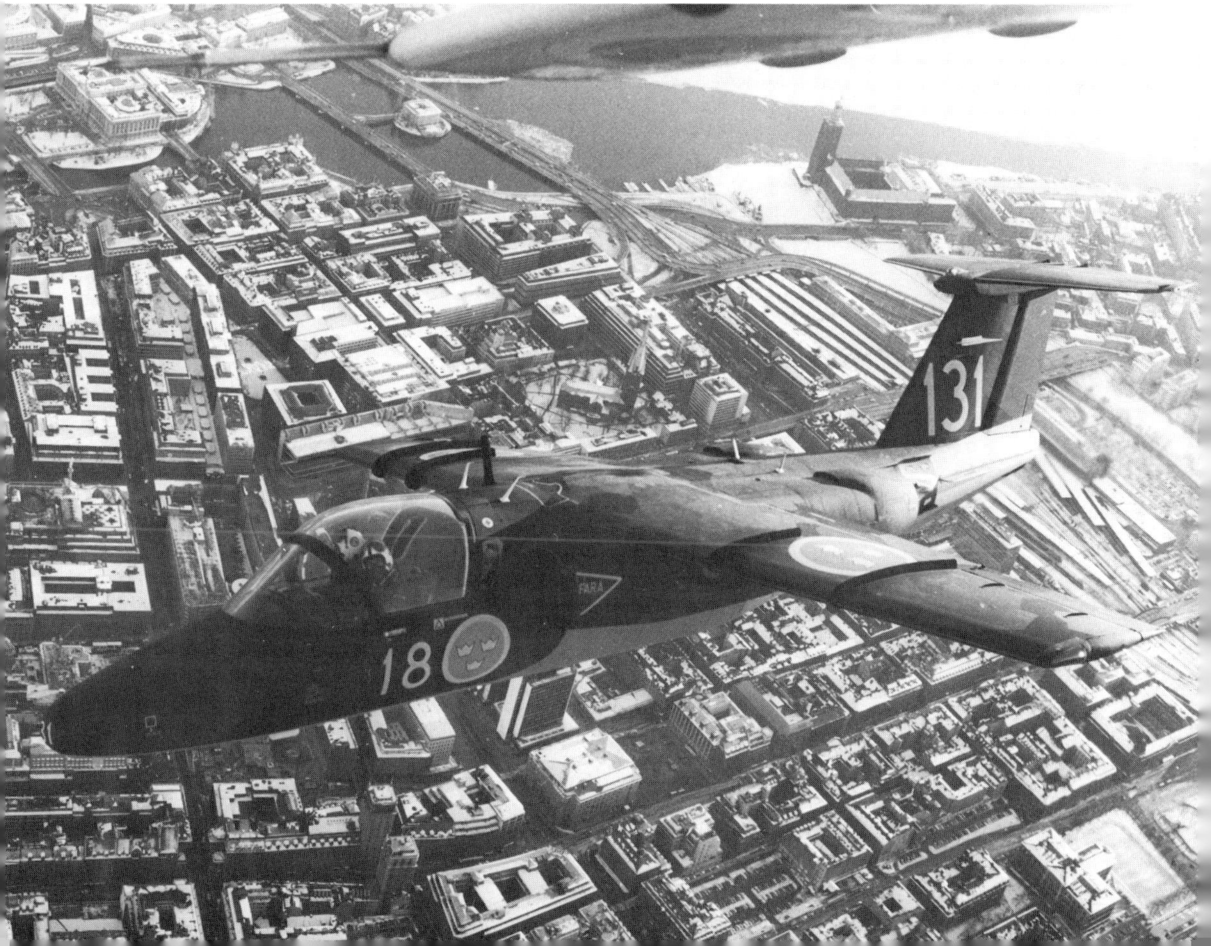
131
18
FARA

Above:
Typifying the continuing use of foreign-designed aircraft, Sweden's entire fleet of C-130 Hercules transports, designated Tp.84, line up at F7 wing at Satenas on 8 September 1981. Swedish Air Force

indigenous. The world is too inter-dependent, the resources of any one nation too few . . .

During World War 2, while surrounded by Nazi-occupied territory, cut off from many routine dealings with the outside world, and still embargoed by Roosevelt, Swedish engineers quietly copied the 1,065hp (794.1kW) Pratt & Whitney TWC-3 radial engine for their locally-produced J22 fighter *without having permission to do so*. After the war, a duly apologetic Swedish delegation marched into the Pratt & Whitney works in Hartford, Connecticut to 'set things right' and offer payment for manufacturing the engine without a licence. American products had been 'lifted' by numerous countries during the war but no other had attempted, afterward, to voluntarily reimburse the maker. So impressed were Pratt & Whitney officials with their visitors' sincerity that they retroactively sold permission rights *for a fee of one dollar!* This is often cited as a heart-warming story but it illustrates the near-impossibility of true self-reliance. Foreign technology also contributed to the guns, gunsight, instruments and radio used by the J22.

Outside assistance was needed with components of other wartime aircraft and with every major design which followed, all the way up through the System 32 Lansen fighter.

To return to the modern world, in Viggen's immediate predecessor, the very popular Draken fighter (next chapter), the drive to achieve an 'all-Swedish' airplane again took a strong drubbing. As will be noted, the Draken was not merely a native product but a very original one indeed, no other fighter in the world being quite like it. But self-reliance remained an elusive goal. The Draken's Volvo Flygmotor RM6C turbojet engine which delivers a respectable 12,709lb (7,830kg) of dry thrust is, in fact, nothing more than a licence-built Rolls-Royce RB-146 Avon 300.

Even the Draken's 30mm Aden cannon, mounted on the right wing and stocked with 90 rounds, is of 'foreign' design. The Draken also owes to non-Swedish technology much of its electronics and avionics. General Soderberg's wish 'never to be dependent again' has seen fulfillment to some extent thanks to the remarkable achievements of a dynamic local aircraft industry. But even with Viggen itself, self-reliance would remain a goal, not an achievement. True independence would remain, as it is today and will be tomorrow, an ideal, not at attainment.

3 Draken

Before the Viggen came the Draken. Conceived by SAAB's well-honed design team under Erik Bratt, the System 35 Draken (Dragon) owes its origin to a 1949 *Flygvapen* draft requirement for an interceptor to bulwark Sweden's air defences.

Although no fighter in the world was capable of supersonic flight at the time – the North American F-100 Super Sabre would come a half-decade later in 1954 – the Swedish Air Force had already decided that a fighter with speeds up into the Mach 1.8 range would be necessary to guard against a coming generation of Soviet jet bombers. To these very ambitious requirements were added a need for a fighter which could operate not merely from conventional airfields but (as if in anticipation of the later STOL requirement for Viggen) from straight stretches of prepared roadway 6,560ft (2,000m) long, some of which had been widened to 82ft (25m) while others were only 42ft 8in (13m) across. The new fighter would need to have excellent visibility and ground-handling characteristics and be amenable to temperature and climate extremes.

To carry out their charge, Erik Bratt's design team simply had no choice but to develop a machine which is still among the more advanced and most innovative of its kind. The unique, double-delta Draken put Sweden in the forefront of supersonic development.

Reflecting the thoroughness of the SAAB-Linkoping team, which had flown a prop-driven Safir with the J29 Tunnan's swept wing before the J29 itself ever flew, a mini-sized Draken called the SAAB 210 was put aloft for tests. Since there was little data available on the handling characteristics of delta-winged air-craft and none whatever on the double delta, the SAAB 210 or Lilldraken (Little Dragon), made to 7/10ths scale, was an ideal test platform and provided cost-effective in-flight experience from its first flight on 21 January 1952.

Powered by an Armstrong Siddeley Adder engine developing 1,050lb (476kg) thrust, the SAAB 210 testbed had a disproportionately large cockpit with a bulbous canopy which gave the pilot clear all-round visibility. Since it was intended to test the double-delta planform at lower speed regimes, it had landing gear which was only partially retractable. It was, above all, evidence of Swedish initiative: design teams in other countries didn't create scaled-down flight test models in advance of prototype fighters, but what others did was of no concern.

The double delta planform introduced a sharp 'bend' in the leading edge of what otherwise would have been a conventional delta wing, permitting large useful

Below:
The SAAB 210 experimental test aircraft was a scaled-down replica of the Draken fighter, employed to test Draken's unique double-delta configuration. The test ship was originally flown in the short-nosed configuration seen here. *Saab, via Maj Lennart Berns*

Right:
Modified to a long-nosed shape closer to that of the J35 Draken to follow, the SAAB 210 test craft lands with brake chute. A little-known fact is that this scaled-down test machine was originally christened Draken. It retrospectively became Lilldraken (Little Dragon) when its original name was decided upon for the fighter for which it paved the way. *Saab, via Maj Lennart Berns*

Below:
Shortly after roll-out in October 1955, the brightly-painted SAAB 35 Draken aims its candy-stripe pitot probe towards Nordic skies. An exhaustive flight test programme with its 'mini' predecessor, the SAAB 210 Lilldraken, eased the way for an eventless first flight for this unorthodox fighter. *Saab-Scania*

internal volume at the wing root while the outer part of the wing flared more sharply and became thinner. The double delta configuration facilitated location of the centre of gravity (CG) near the aerodynamic centre of the aircraft. Apart from wind tunnel and other theoretical tests, the double delta was subjected to nearly 700 hours of flight examination in the SAAB 210 testbed and no special problems were encountered.

Flygplan 35, the Draken, was rolled out from SAAB's Linkoping factory already the beneficiary of an exhaustive flight test programme before it had even flown. On 25 October 1955, piloted by engineer Bengt Olov who was for many years SAAB's chief test pilot, the Draken made its uneventful first flight. Three prototype Drakens were powered by imported Rolls-Royce Avon engines, subsequent machines to be powered by the 12,690lb (5,756kg) static thrust Volvo Flygmotor RM6C turbojet, the Swedish-built version of the Rolls-Royce RB146 Avon 300. This engine, in its Swedish variant with an afterburner more powerful than any used with the British design, produces 16,800lb (7,620kg) static thrust with afterburning. Although not 'Swedish' as noted in the previous chapter, the engine has been refined through Swedish manufacture and has proven remarkably reliable and trouble-free. Internal fuel is carried by the Draken in the inner wings (the stowage space afforded by the double delta planform) and fuselage bladder tanks. Total internal fuel capacity is 880Imp gal (4,000l). External tanks under the fuselage and wings can increase fuel capacity to 1,980Imp gal (9,000l). Additional internal fuel tanks can be fitted in place of guns for ferrying purposes. Although the Saab 210 Lilldraken could adjust its centre of gravity by pumping fuel between tanks, the Draken has a more limited cross-pumping feature. The production Draken fighter has a single point pressure fuel system with a capacity of 185Imp gal (840l)/min.

Draken Variants

Following initial flight tests which included good performance in supersonic speed ranges, the Draken was ordered into production in 1956. The first production J35A Draken fighter made its first flight on 15 February 1958, and deliveries to air force wing F13 (F for *Flygflottilj*) at Norrkoping began in March 1960. Since then, the following variants of the Draken have been built:

J35A The initial production version was powered by the Svenska Flygmotor (later Volvo Flygmotor) RM6B, the locally built version of the Avon 200, with Swedish-developed afterburner, giving 11,000lb (5,000kg) static thrust without afterburner and 15,200lb (6,900kg) static thrust with. This model had SAAB S6 fire-control equipment and a Lear autopilot. Not fully successful at first, the J35A was restricted to rear hemisphere attacks in its interceptor role and was soon superceded. An aerodynamically improved and lengthened rear fuselage, with dual tail-wheel unit, was retrospectively applied to this version. Ninety aircraft of the J35A variant were produced, with FV serial numbers 35001 to 35090. Some were later converted to the J35B standard, others into Sk.35C trainers.

J35B This second interceptor variant of the Draken was equipped with a SAAB S7 collision-course fire control system and with data-link equipment for the STRIL-60 air defence system. The improved and lengthened rear fuselage was included on this model from the beginning. The J35B first flew on 29 November 1959 and was soon in service with F18 wing at Tullinge, near Stockholm, among other units,

Below:
Being taxied at Linkoping by SAAB test pilot Bengt Olov, the number one SAAB 35 Draken is readied for its October 1955 maiden flight. The clear, unbroken windscreen was to be a trademark of Draken and of the Viggen to follow. *Saab, via Maj Lennart Berns*

Above:
This early shot of two J35F Draken interceptors in natural metal emphasises the double-delta planform of the popular fighter. These machines carry two each of the Rb.28, a Swedish-built version of the Hughes AIM-4 Falcon air-to-air missile. *Saab-Scania*

where it served with the colourful 'Acro Delta' flight demonstration team. Eighty-nine aircraft of the J35B type were converted or built, with FV numbers 35201 to 35289. Many were later brought up to the J35D standard.

Sk.35C The two-seat trainer version (Sk for *skolan*, or school) made its first flight on 30 December 1959. The Sk.35C retained the short tailcone of the initial J35A model and carried no armament. Used primarily at F16 wing at Uppsala to train Draken pilots, the Sk.35C positioned the flight instructor behind the student in tandem and provided him with a periscope for improved visibility, a feature which would appear again in the two-seat Viggen. Twenty-five aircraft of the Sk.35C variant were manufactured, with FV serial numbers 35801 to 35825.

J35D The J35D interceptor aircraft was the first to introduce the more powerful RM6C engine, described elsewhere in this chapter as fundamental to the Draken type. First flown on 27 December 1960, this model also introduced the SAAB FH5 automatic pilot and had increased fuel capacity. This was the first model of the Draken to have the SAAB RS-35 rocket-propelled ejection seat for pilot escape at zero speed/zero altitude, if necessary. Ninety-three interceptors of the J35D standard were converted or built, and were assigned FV serial numbers 35301 to 35393.

S35E The *spanning*, or reconnaissance, version of the Draken was based on the J35D interceptor design but had seven camera systems for photo-taking at varying speeds and heights. Equipment includes camera windows fitted with defrosting capability, a nose which slides forward for easy maintenance access to the cameras, a new Swedish-designed camera sight, and the zero/zero ejection seat. First flown on 27 June 1963 and in service since 1965, the S35E variant is unarmed. Sixty of these aircraft were built, assigned FV serial numbers 35901 to 35960.

J35F The definitive interceptor Draken, the J35F was built in two types which have an important recognition difference: J35F-1 *(Filip Ett)* aircraft have no infra-red scanner under the nose radome and are identified by odd-numbered unit codes; J35F-2 *(Filip Tva)* aircraft have the IR scanner and are identified by even unit codes. The J35F aircraft have an improved SAAB S7B collision-course fire-control system and improved Ericsson PS-01/A intercept radar.

The J35F interceptor has a maximum speed in 'clean' condition of 1,320mph (2,125km/hr) or Mach 2.0 at 36,090ft (11,000m). Empty weight of the J35F is 16,369lb (7,425kg) and maximum take-off weight is about 28,000lb (12,700kg). A relatively compact aircraft, the J35F has a wing span of 30ft 10in (9.40m), an overall length of 50ft 4½in (15.35m), a height of 12ft 9in (3.89m) and a wing area of 529.6sq ft (49.20m).

A Successful Fighter

The J35F Draken is armed with one 30mm Aden M/55 cannon with 90 rounds in its right wing, the left-wing cannon found on earlier models having been deleted to make space for avionics gear. Its normal missile complement is two SAAB-produced Rb.27 Falcon HM-55 (radar-guided) and two Rb.28 Falcon HM-58

Top:
A December 1960 view of the Draken production line at SAAB's Linkoping plant. Once the Draken programme was embarked upon, production reached a peak of 55 airframes per year. *Saab-Scania*

Above:
Denmark is the only NATO country to operate Swedish fighters in squadron service. Aircraft 01 was the first Draken in the Danish programme. *Saab-Scania*

(infra-red) air-to-air missiles. The 'punch' of the J35F Draken – which can be employed in the air-to-ground role – is enhanced by 11 attachment points for external stores: three 110kg and one 500kg under each wing and three 500kg under the fuselage. A typical ordnance store could be 18 13.5cm Bofors air-to-ground rockets, bombs or fuel tanks.

The J35F is by far the most numerous of Draken variants, no fewer than 208 having been manufactured with FV serial numbers 35401 to 35608.

At one time the ubiquitous J35F equipped no fewer than eight fighter wings (F1, F3, F10, F12, F13, F16, F17 and F21). A service life extension program undertaken in 1981 will refurbish and update 70 J35Fs to serve with three *Flygvapen* wings perhaps to the end of the century although, as will be noted shortly, this programme was threatened by a funding debate.

Flying the Draken
Piloting the Draken is a special experience, remembered fondly even by pilots who have since moved on to the more advanced Viggen. The pilot enters the aircraft by climbing a metal ladder designed to hang from the left side of the cockpit. He sits slightly nose-high taxying, his view unobstructed through the clear windshield. Take-off is normally made with afterburner, and the Draken is extremely

responsive to the pilot's touch at the controls during lift-off, while being vectored toward the target, and in combat. Pilots are cautioned not to be too trusting and not to take liberties with manoeuvre restrictions: the Draken, once again, does not forgive.

A major with 10 years' service and about 800 hours in Drakens remarked upon an intercept of a Soviet Tupoleve Tu-26M ('Backfire') bomber over the Baltic Sea, on the condition that he not be identified. 'The Draken is very receptive to sudden, minor course adjustments but you have to watch for a stall indication every second. This responsiveness was a great help when I found myself coming up behind that Backfire from his left rear. He wanted to "jink" out of my way and I simply stayed with him. It was a brilliant winter afternoon, the sun glinting off his wings and fuselage, and I felt I could have shot him down if I'd had to.'

One of the very few foreigners ever to fly a Draken was Capt Michael P. Curphey of the US Air Force, who served in Stockholm in an exchange capacity in 1969. 'I was given an orientation flight in an Sk.35C as a kind of "reward" for three years in Sweden working with *Flygvapen* officers', says Curphey. During the one-hour flight from F16 wing at Uppsala, during which time Curphey occupied the student pilot's front-seat position, his instructor pilot treated him to a demonstration of the well-known Swedish ability to fly at treetop level. 'We were out over the Stockholm archipelago, the cluster of islands between the capital and the Baltic Sea. We kept flying past these small islands which had *stugas* (vacation homes) on them. The astonishing thing was, I would look out and see somebody's house going past to our side and *above* me!'

Foreign Drakens

Flygplan 35 – the Draken – was exported to Denmark and Finland. Had its country not been a politically neutral state, it might well have been sold elsewhere. Indeed such an advanced fighter type produced by, say, a British or American manufacturer would certainly have been assured of strong export prospects!

In NATO Denmark, squadrons Esk 725 and Esk 729, both stationed at Karup, are the users of the Draken. Each of the two squadrons operates all three Danish versions: The F-35 is the ground attack fighter variant, also known by the company designation SAAB 35XD; the RF-35 is similar but is a

Left:
Pilots in standard flying garb surround an operational, camouflaged J35F-2 Draken. The flight suit worn by Swedish fighter pilots is almost identical to that used by the RAF; a life-vest is mandatory because of the frequency of over-water missions, especially in the Baltic where Soviet aircraft prowl. *Swedish Air Force*

Top:
Camouflaged J35F Draken interceptors of F13 wing at Norrkoping, which has subsequently become the first unit to use the JA37 interceptor Viggen. The white circle on upper wing of foreground aircraft is for recognition purposes on training missions. *Robert Lofberg*

Above:
An early photo taken prior to adoption of camouflage, showing a J35 Draken of F13 wing at Norrkoping landing with parachute brake. *Saab-Scania*

reconnaissance variant with the same provision for cameras as the S35E; and the TF-35 is Denmark's two-seat trainer variant.

The first Danish machine was delivered on 29 April 1970. Denmark has 20 F-35 aircraft with military serials A-001 to A-020 (constructor's numbers 351001 to 351120). Of these, four F-35 airplanes (A-003, A-013, A-015 and A-016) have been written off in accidents. Denmark also operates 20 RF-35 Drakens, with military serials AR-101 to AR-120 (c/ns 351101 to 351120), of which two (AR-101 and AR-103) have been write-offs. Eleven TF-35 two-seat Drakens were also obtained, which were to bear military serials AT-151 to AT-161 (c/ns 351151 to 351161).

The non-aligned Finnish Air force, or *Ilmavoimat*, with its strange mixture of Western and Soviet equipment, flies no fewer than four separate variants of the Draken. The J35BS is Finland's equivalent of the J35B interceptor but with radar deleted and minor changes. It is used for training only. The J35F is identical to the Swedish J35F and comes second-hand from Swedish inventory as does the two-seat J35C, the Finnish equivalent of the Sk.35C. The J35S is a simplified export model for Finland. Finland's No 11 Squadron operates MiG-21s at Kuopio-Rissala and Drakens at Rovaneimi! In 1983 it was announced that Finland would obtain about 20 ex-Swedish Air Force J35Ds valued at FMk 195million (£29million) to equip a second fighter squadron at Satakunta in southern Finland.

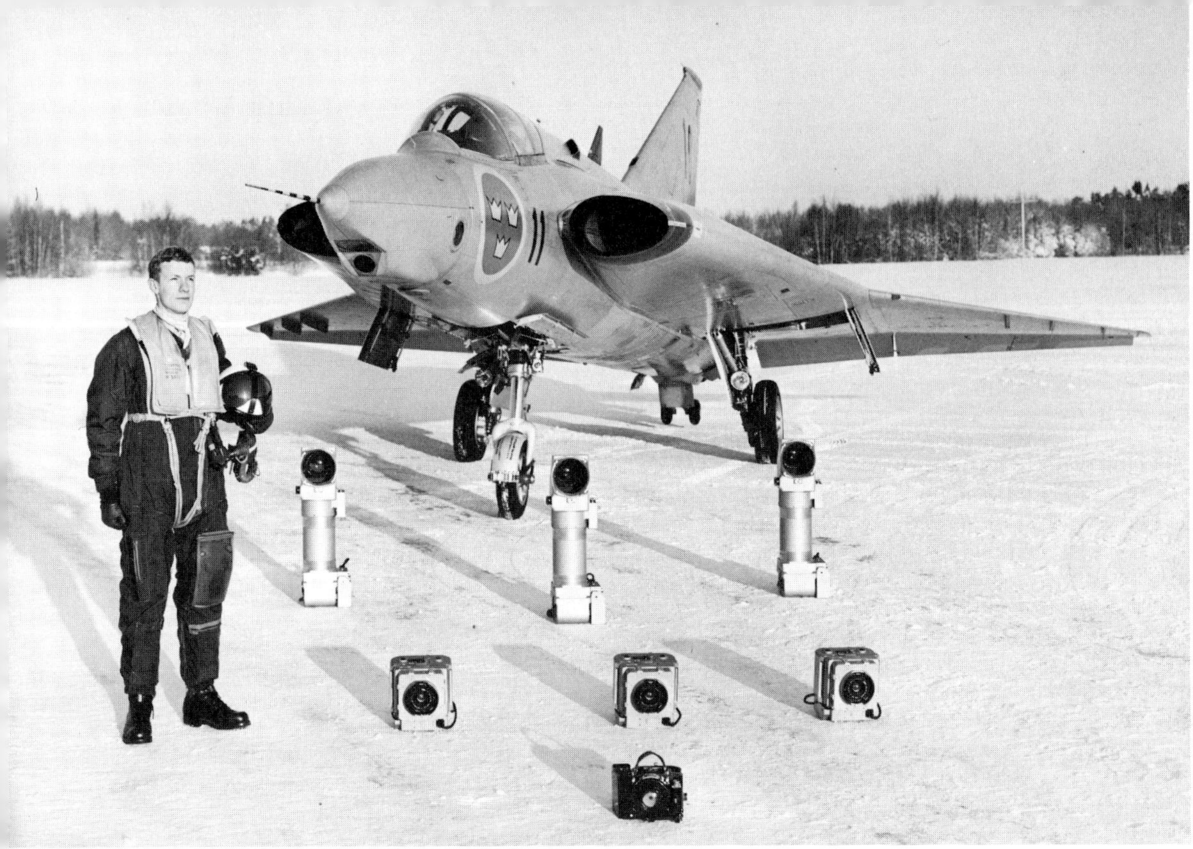

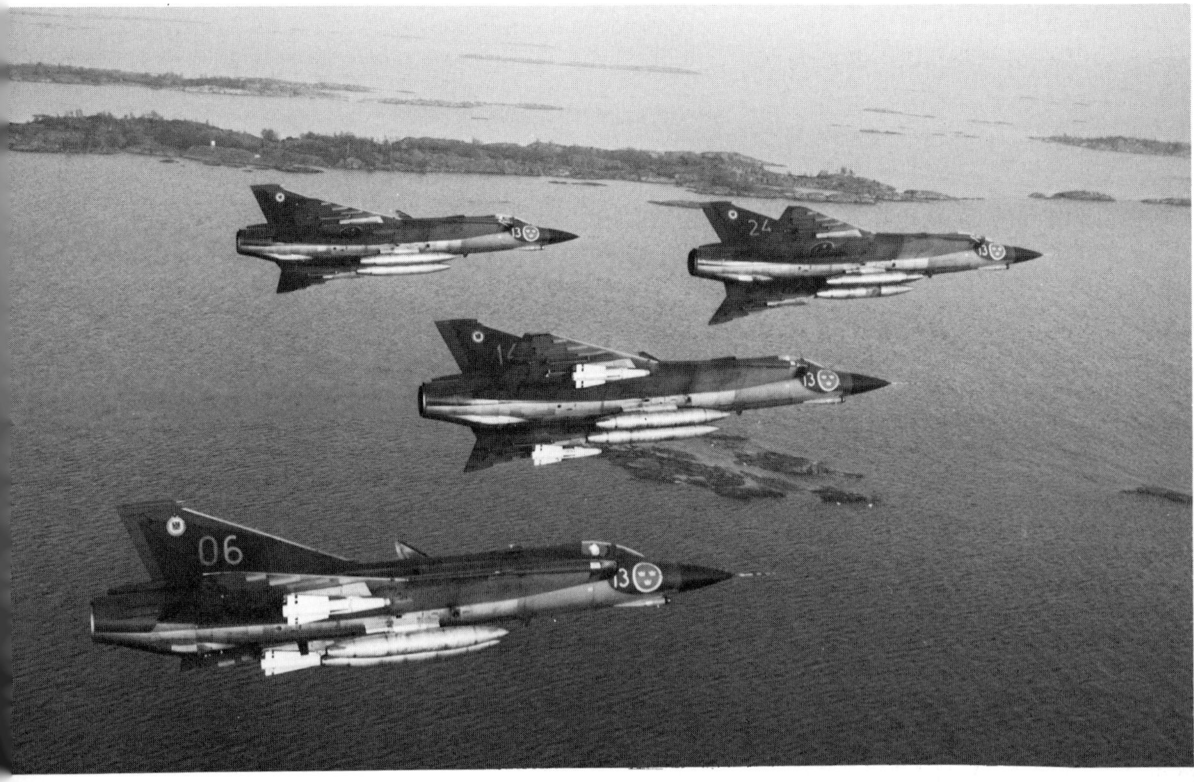
13
24
13
14
13
06
13

Far left:
The S35E reconnaissance Draken was based on the J35D interceptor design but had seven camera systems, displayed here in front of the aircraft. The unarmed S35E has been in service since 1965. *Saab-Scania*

Below left:
En route eastward over the Stockholm archipelago heading for the Baltic, four J35F-2 Drakens from F13 wing at Norrkoping carry drop tanks and Falcon missiles. By 1976 all Drakens in operational service were camouflaged. *Saab-Scania*

Left:
A pilot clambers aboard this camouflaged Draken. The infra-red (IR) probe beneath the nose radar identifies this as a J35F-2 model. *Swedish Air Force*

Below:
This early view shows J35A Drakens in pleasant sunshine on the line at Norrkoping. *Saab-Scania*

Above:
The Draken looks interesting from any angle. This rearward view shows the size of the aircraft relative to its pilot. Swedish Air Force

Right:
A Sk.37 Viggen of F15 wing at Soderhamn, with the centreline external fuel store that is required in all flights by the two-seat conversion trainer. An auxiliary power unit (APU) is used in nearly all Viggen operations. *Herman J. Sixma*

The Draken's Future

The Draken remains uncommonly popular with pilots and aviation buffs. All of its variants have served well, with minimal teething problems, despite the uniqueness of the double-delta planform. The Draken has not, however, earned a reputation as a forgiving airplane and is by no means easy to master. Many of the problems it has encountered – as in 1983, when one squadron of Colonel Bert Stenfeldt's F21 wing at Lulea lost four Drakens in four months – have been officially blamed on inadequate training rather than faulty design. In a 15 February 1983 accident, a pilot ejected after 'super-stalling' his double-delta craft.

One of the saddest of Draken losses occurred on 4 April 1984 when a J35D of F4 wing at Ostersund flown by Lieutenant Colonel Mats Tjarn piled-up on a frozen lake while descending on approach to the base. The Draken touched down on the lake, gear down, amid an ice storm which seriously hampered visibility. It began to burn when a drop tank was ripped off and set ablaze. A private plane flown by former Draken pilot Erik Wollo landed on the lake to help. Wollo struggled desperately to release Tjarn, who was trapped in the burning fighter for more than 20 minutes. Tjarn died in a hospital.

Still, overall losses of the Draken compare favourably with those of other modern fighter types. The airframe has considerable 'wear' left on it. Thus it was surprising that for a time in 1984 the future of Sweden's Draken squadrons seemed in doubt.

The Swedish Air Force had always planned to retire most of its J35D and J35F Drakens as JA37 interceptor Viggens were delivered but had intended to retain three Draken squadrons with F10 wing at Angelholm for an extended period. The future of these squadrons seemed emperiled when a typical funding debate arose in early 1984. The *rikstag*, or Parliament, was divided over the amount of supplemental defence expenditure to be added to five year projections made in 1982 – a rift typical of the political disputes which often intervene in military decisions. General Lennart Ljung, supreme commander of Swedish forces, wanted an additional 2.5 billion kronor (£225million) to keep Drakens flying. For a time, with Prime Minister Olof Palme's Social Democratic government giving priority to the JA37 Viggen, it seemed Ljung wouldn't get the money.

A compromise on a figure of 2.2 billion kronor (£195million), which left funds available for other priority needs – the JA37, the JAS39 Gripen and the AIM-9L Sidewinder – culminated in a 14 March 1984 agreement which assured the future of the Draken. A major retrofit programme planned for the Drakens had to be abandoned, however, and only improvements related to aircraft safety are now planned.

AIM-9L will give Drakens and Viggens much-enhanced dog-fight capability. Proven in the Falklands War where Sea Harrier pilots used it effectively, it is the first infra-red (IR) 'fire-and-forget' missile capable of homing on an enemy's exhaust heat, even in a frontal, head-on engagement.

Because many Drakens were converted to a later standard from an earlier one, and because official figures are not available, it remains unclear precisely how many examples of this attractive fighter aircraft were built. It would appear that 545 machines were produced for the Swedish Air Force, 51 for Denmark, and 12 for Finland (not counting Finland's ex-Swedish

Top:
A J35F-2 with drop tanks from F13 wing banks over the Swedish countryside. *Robert Lofberg*

Above:
Towbar attached to its nosewheel, a lonely J35 Draken interceptor of F3 wing at Malmslatt sits in the elements, awaiting display at one of the periodic International Air Tattoo shows at RAF Greenham Common, England.
Swedish Air Force via Bryan Philpott

aircraft) for a total of 608 Drakens built. Like the Viggen to follow, the Draken was offered to a number of other potential export customers. Switzerland, though it never purchased the aircraft, got as far as having an export model designed to meet its own particular specifications.

Given the other obvious qualities of the design, Draken now seems likely to remain in service at least until 1989.

4 Design, Development and Flight Test

As long ago as May 1952, as already noted, the funding and inspiration were available for Erik Bratt's design team to stretch its talents, to innovate, and to envision the fighter aircraft which would serve from the early 1970s through to the end of the century. Bratt *wanted* innovation. He is reported to have told one aerodynamicist to disregard all day-to-day obligations, sit in a corner, and *think*.

Self-sufficiency, in the true sense, remained an elusive goal. In the 1950s and 1960s the de Havilland Vampire, Mosquito NF19, Venom and Hawker Hunter – given designations J28, J30, J33 and J34 respectively – all served in Swedish fighter squadrons contemporaneously with SAAB's J29 Tunnan, J32 Lansen and J35 Draken. Foreign types such as the Lockheed C-130H Hercules, designated Tp.84, were to perform in key functions. But the country's principal fighter, and its biggest, costliest and most controversial defence project, would emerge from Bratt's SAAB think-tank.

From 1952 to 1958, the largely anonymous and rather sequestered SAAB designers made exhaustive investigations aimed at finding a replacement for the System 35 Draken which, it bears repeating, did not even fly until October 1955. More than 100 proposals were studied, one of these – System 36, what would have been called *Flygplan* 36 or the J36 – being carried well into the design stage before being cancelled. Recurring debate over defence budgets, and over the high cost of producing an indigenous design rather than purchasing a foreign aircraft, seems to have killed the J36 even before this fighter type's design configuration could be decided upon.

The period 1958-61 saw the first really serious studies which were to lead to the Viggen.

Modifications of the Draken, among other proposals, were investigated. (No ground attack variant of the Draken had been developed. The future Viggen, although multi-purpose, would begin its service life in the ground-attack role.) SAAB designers explored both single and two-seat configurations but leaned

Above:
An early concept for the Viggen, believed dated 1963, shows the basic design established, but with air intakes and vertical tail different from those adopted for the actual aircraft. A ventral fin for stability in manoeuvring was discarded in the final Viggen design. *Saab-Scania*

Overleaf:

Top left:
JA37 Jaaktviggen interceptor of F17 wing at Ronneby in standard forest camouflage, in a 45° climb. *Saab-Scania*

Bottom left:
This AJ37 Viggen ground attack aircraft of F15 wing at Soderhamn is seen in the natural metal paint scheme which was largely abandoned in the 1970s.
Herman J. Sixma

Top right:
JA37 Viggen second-generation interceptor of F13 wing at Norrkoping, in the snow-white camouflage scheme which was tested but not adopted operationally. *Saab-Scania*

Bottom right:
In forest green-brown camouflage, an AJ37 Viggen of F13 wing of Norrkoping is caught in mid-air over the runway's end. *Herman J. Sixma*

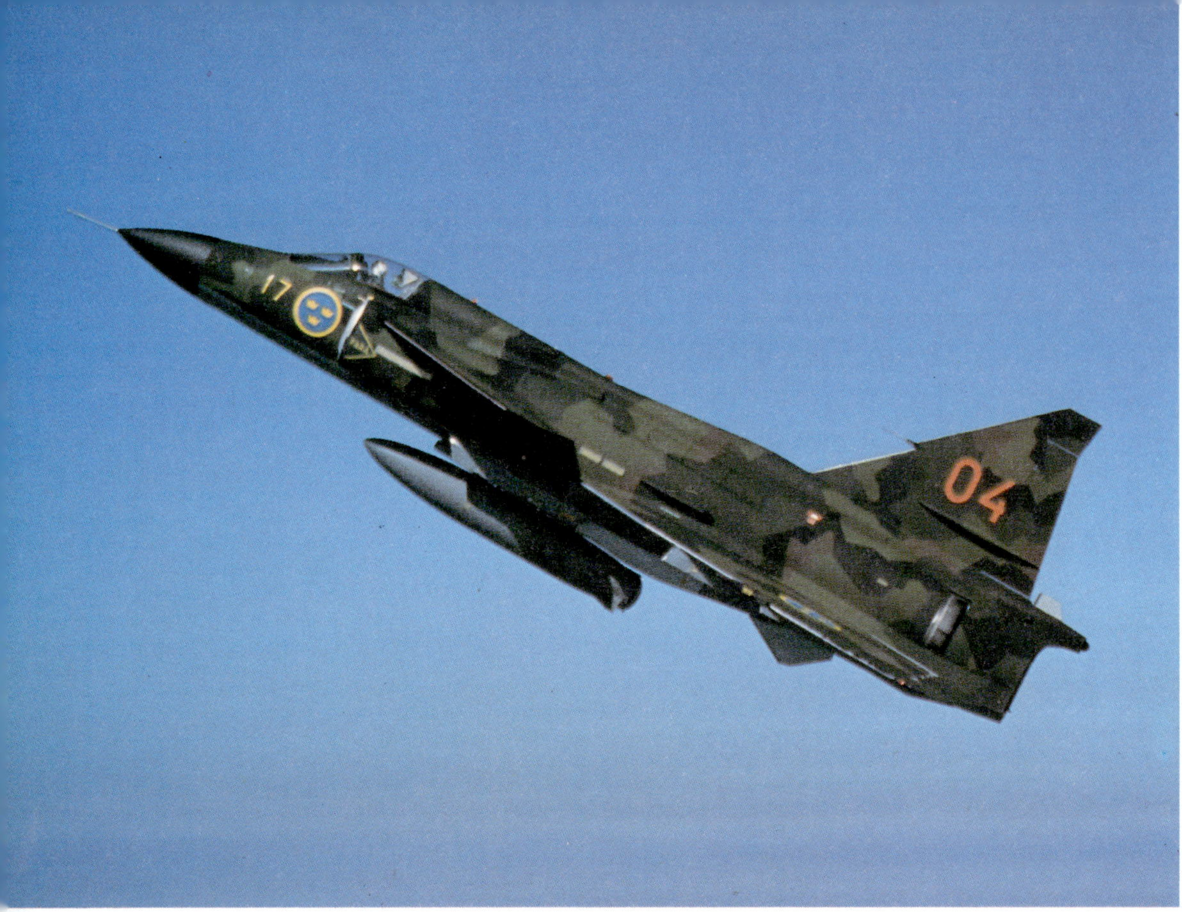
17
04

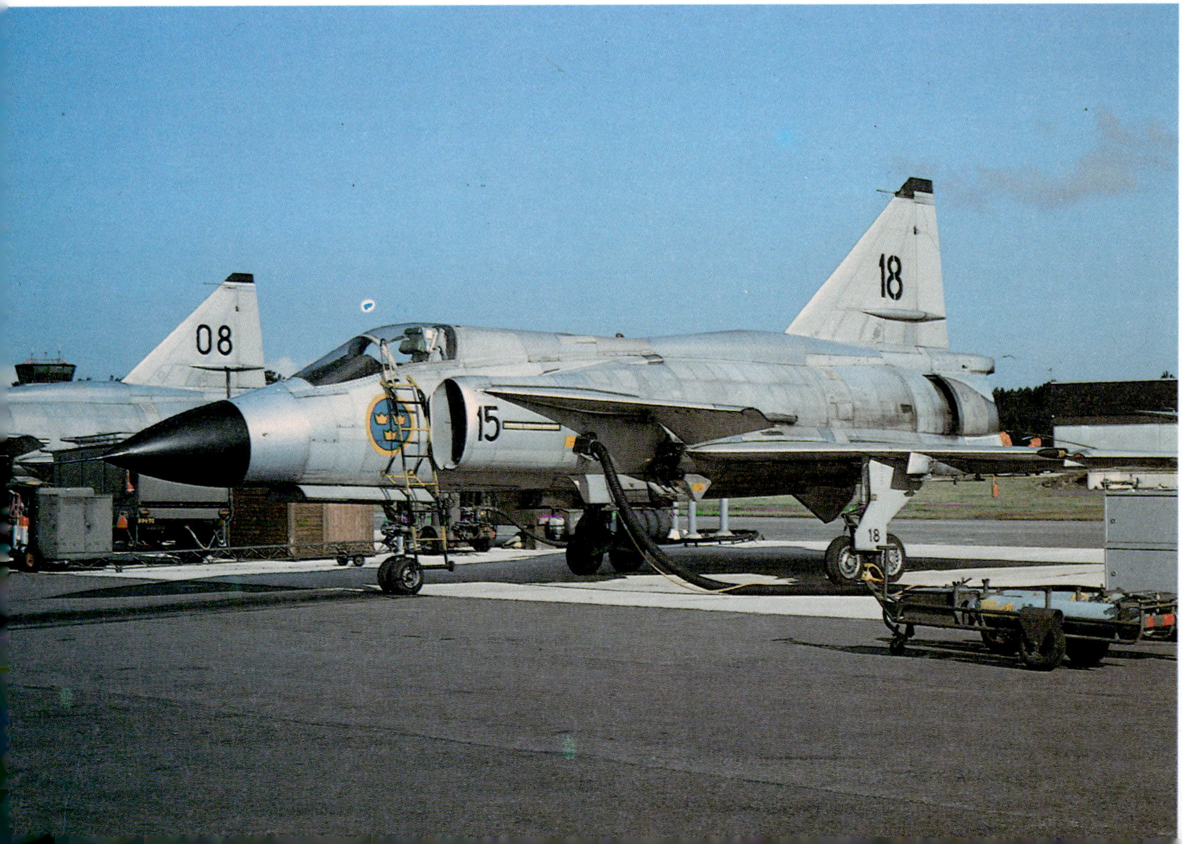
08
18
15
18

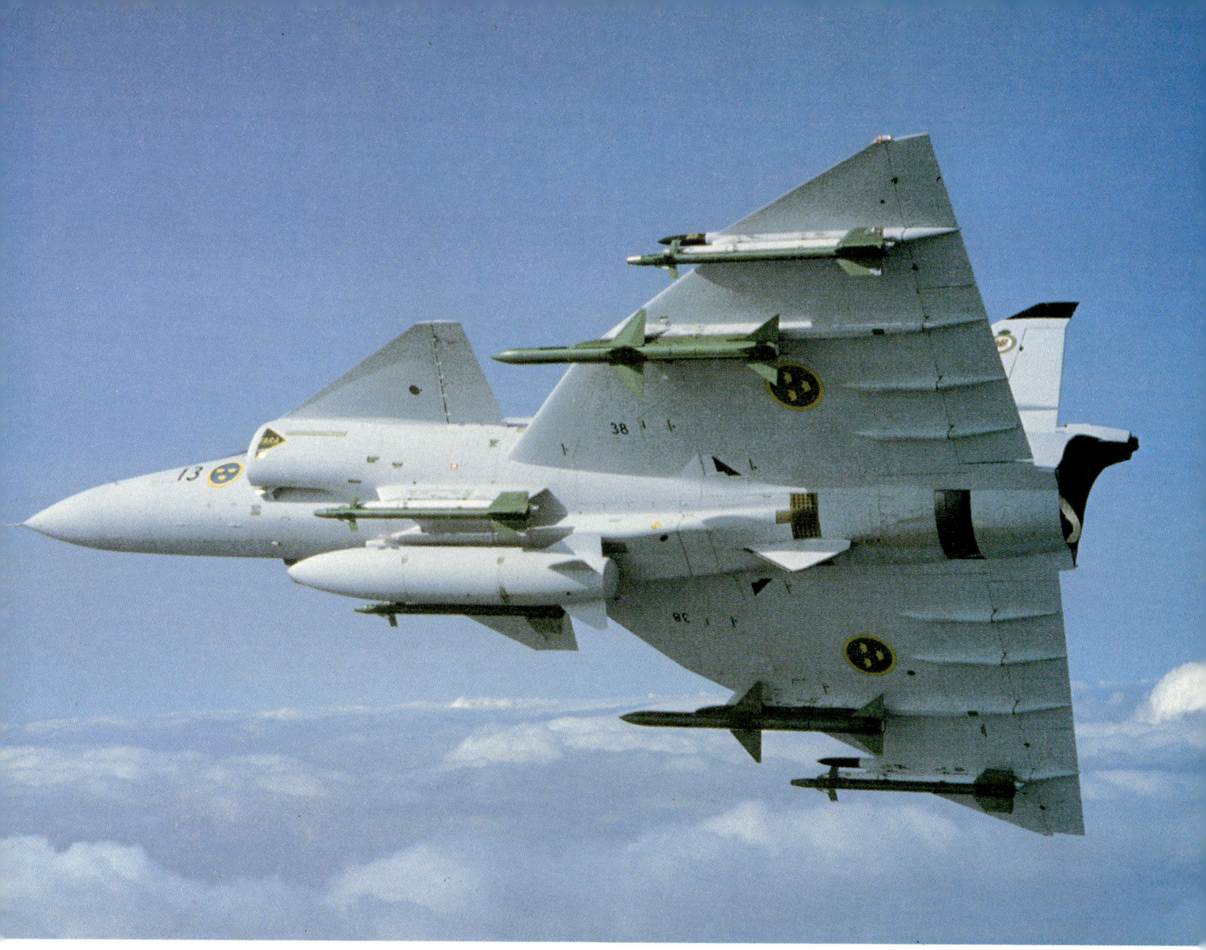
13
38

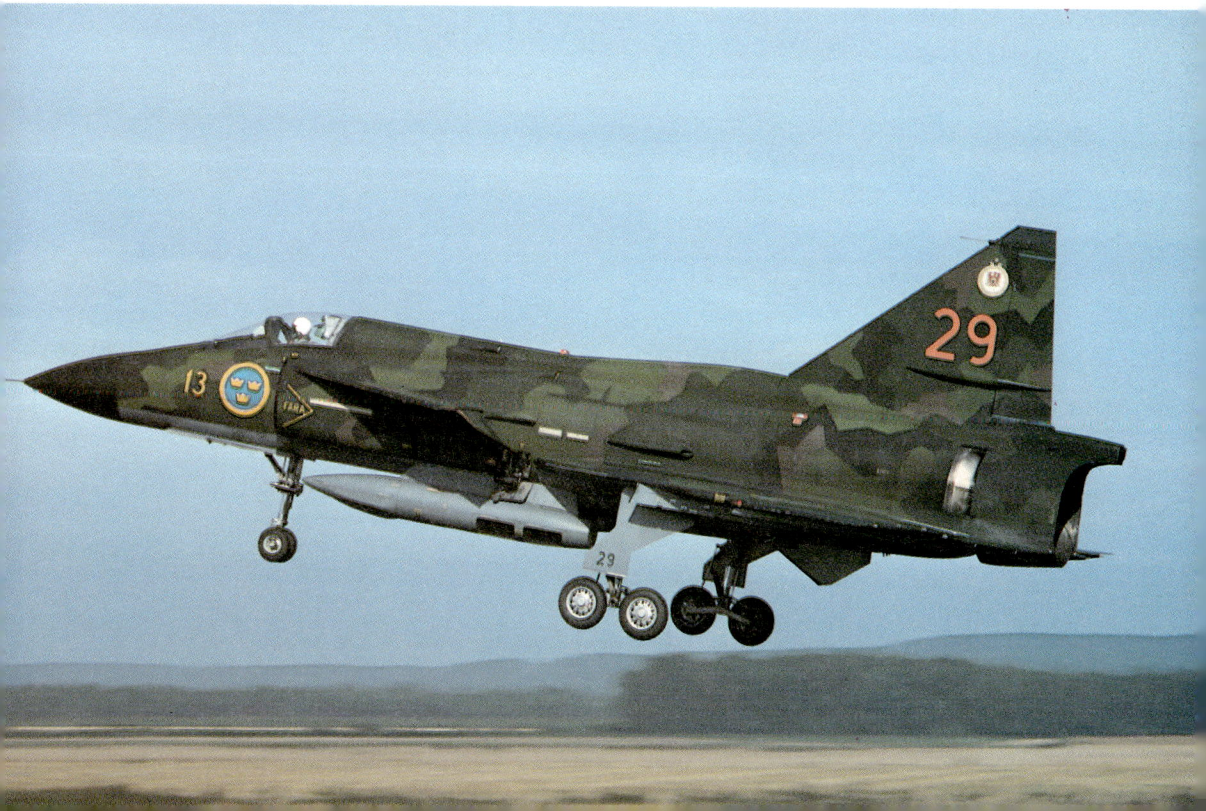
13
29
29

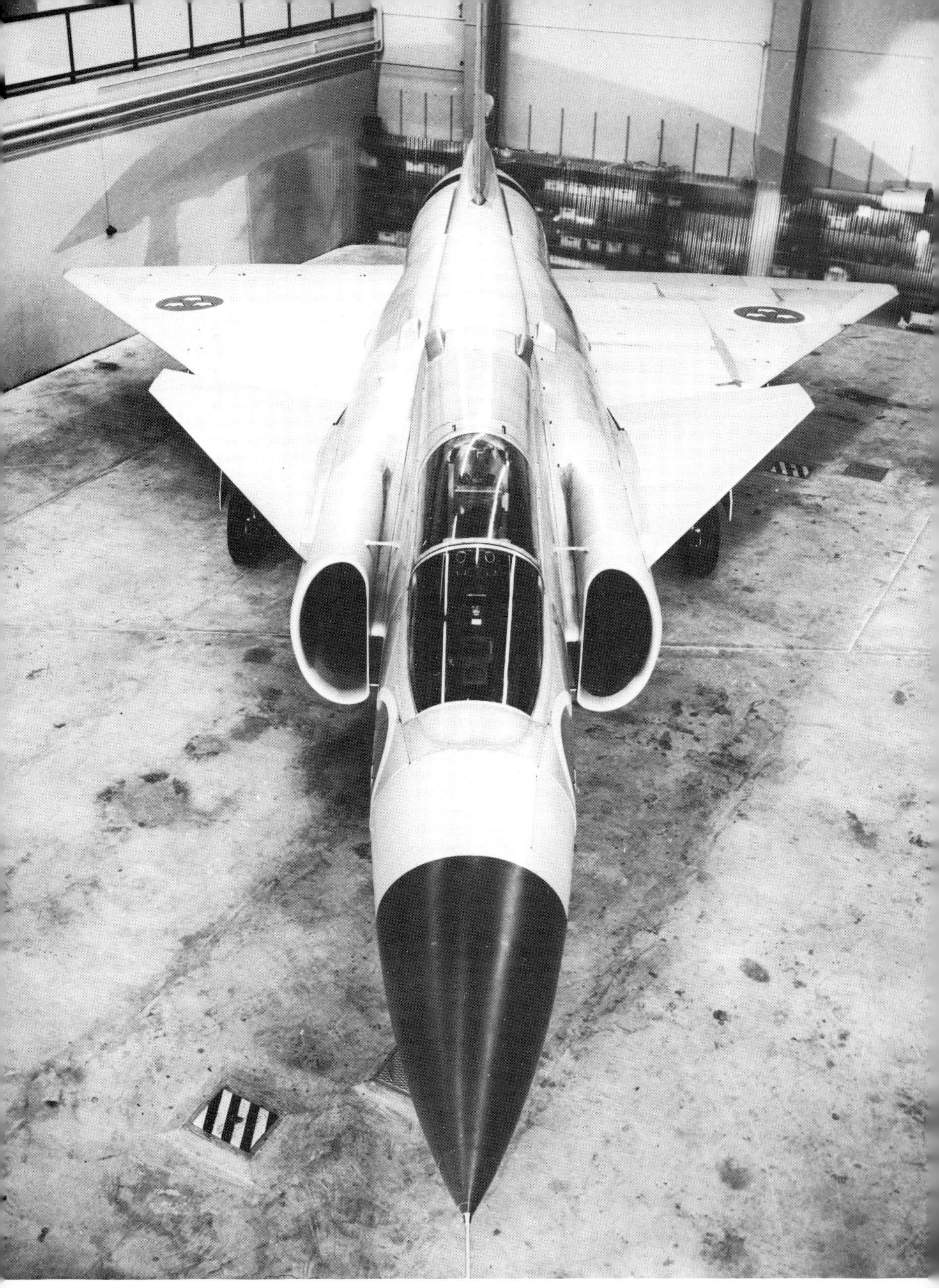

heavily toward an aircraft that could be handled by one man, perhaps reflecting the Stockholm air staff's operational doctrine, perhaps an obvious consequence of the country's small population, perhaps neither.

It is interesting to note that, at the time, things gave the appearance of moving in the other direction. When the two-seat F-4C Phantom replaced the one-man F-84F Thunderstreak with a US Air Force unit in Florida in December 1962, a newspaper reported that, 'Never again will a one-seater fighter be used by a major air arm.' Washington's excitement over the two-man TFX design, which became the F-111, seemed to confirm this, as did the RAF's development of the two-seat Buccaneer.

In Sweden, as in all countries, there were stops and starts for political, fiscal and technical reasons. There were some people, as there had been in London back in the mid-1950s, who argued that advances in the missile field made it sheer folly to obtain any airplane at all. Some public men insisted that Sweden's tradition of neutrality based on strong defense was outdated, that the nation should change to the Swiss model of neutrality with but a token defence. To these men, it wasn't missile technology but politics that seemed to make a multi-million kronor jet fighter unnecessary. The issues – cost, politics, purpose – became confused, but inside Bratt's design shop, hard work continued, insulated from the debate, with certain clear goals always in sight. One of the goals by now, from the designers' viewpoint, was that any future fighter would have an ergonometric (man-efficient) cockpit where comfort, convenience and computerised systems would substitute for any second crew member. Discussion on crew size would continue but the engineers felt that, except in a possible training variant, it would be, it had to be, a single-seater; if indeed it ever came to be at all. 'At times we weren't sure', says Rolf Westerberg, a retired Swedish Air Force officer. 'There was so much debate, we didn't think we'd have any airplane at all.'

A Swedish Air Force directive of February 1961

Left:
From above, this full-scale mock-up of Viggen-to-be SAAB's Linkoping plant reveals a sweepback of the canard foreplane which did not survive in the actual airplane. The wing leading edge design would also change before Viggen roll-out. *Saab*

Below:
Known as 37-0, a pun on its non-existent serial number, the mock-up of the future Viggen (like the fighter plane to follow) required an external ladder for pilot entry. This mock-up appears to have a radar warning and homing system (RHAWS) in a pod beneath right outer wing; on the real Viggen, RHAWS protruded from the wing leading edge. *Saab*

Overleaf:
Left:
No.37954, a SF37 Viggen photo-reconnaissance aircraft of F17 wing, at Ronneby in 1984. *Peter R. Foster*

Right:
A JA37 Jaktviggen interceptor awaits a mission at F13 wing, Norrkoping; it is wearing one variation on the low-visibility grey camouflage for air combat manoeuvring, although this is not the paint scheme adopted as standard. *Chris Pocock*

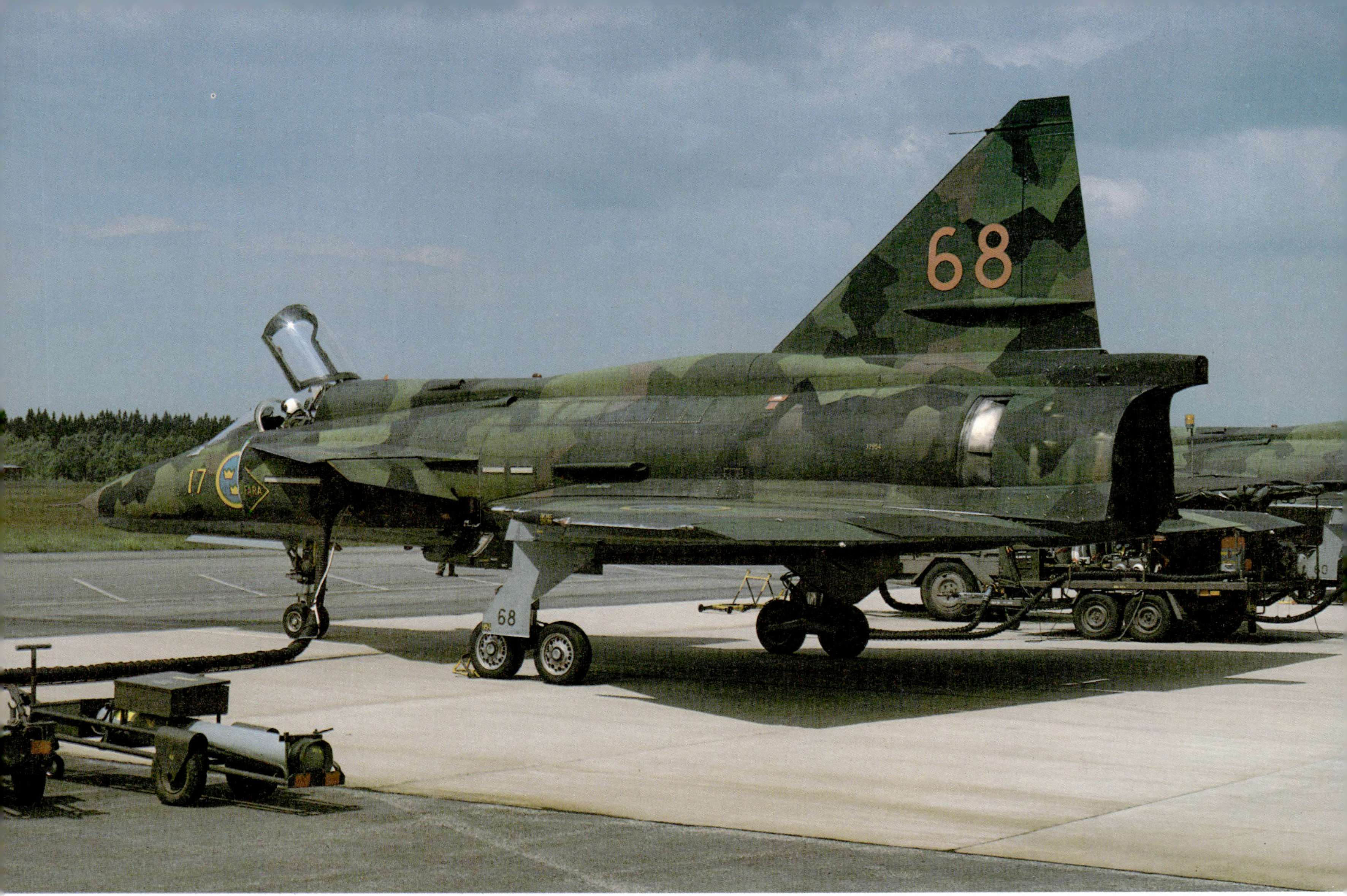
68
68
17

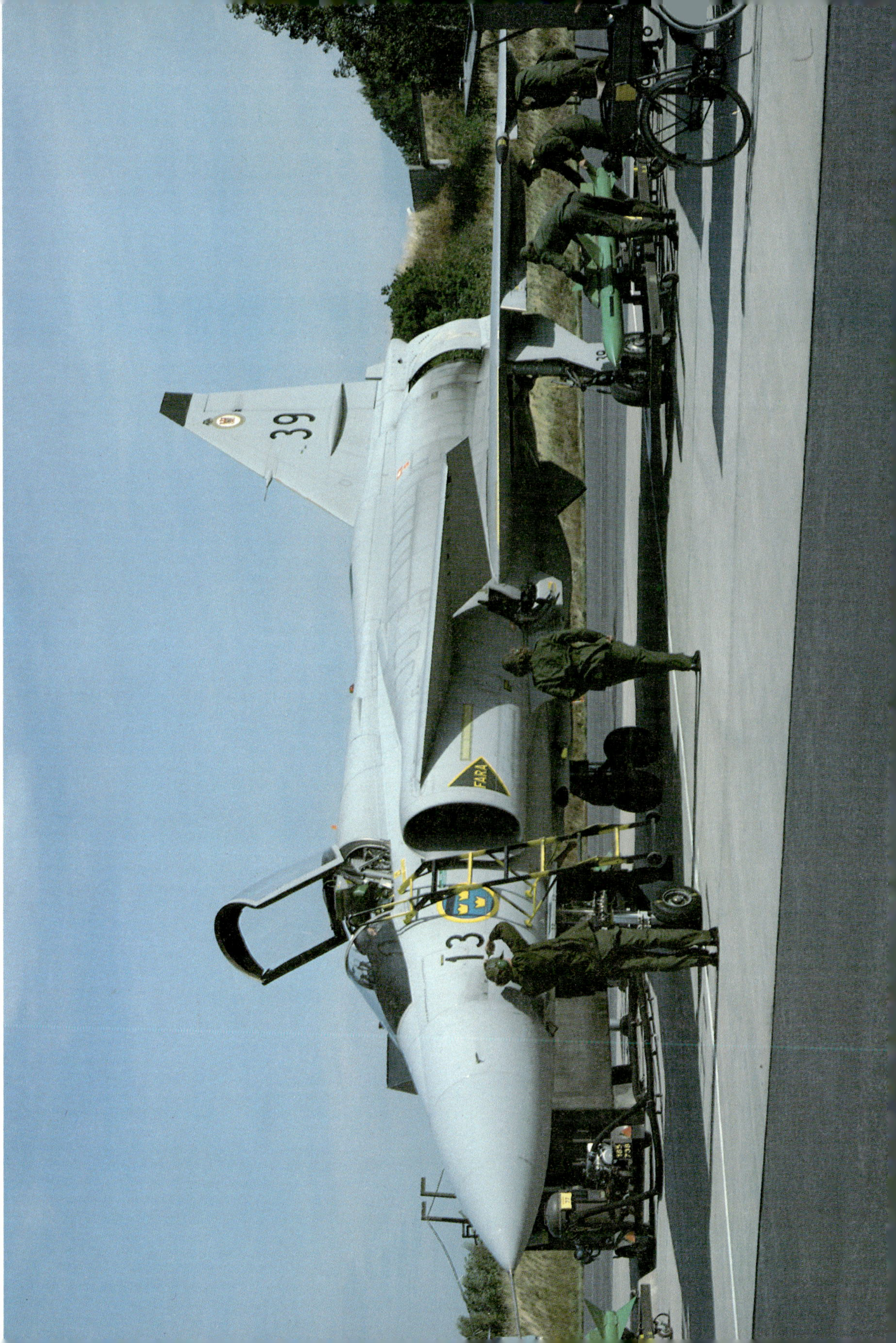
39
13
FARA

Left:
From straight front, with hangar background airbrushed out, Viggen mock-up 37-0 reveals some detail – including nose wheel landing lights – not found in the final aircraft. *Saab*

Below:
Viggen mock-up 37-0 had a straight wing leading edge. Beginning with test aircraft 37-4, the outer leading edge protruded farther forward than the inner, producing the characteristic 'sawtooth' look. *Saab*

Right:
Eric Dahlstrom, the test pilot who made the first flight in a Viggen. *Saab-Scania*

Below right:
The first Viggen: 37-1, shortly after its first flight with Erik Dahlstrom at the controls, pulls above the clouds on a 1967 test hop. *Saab-Scania*

formally gave SAAB responsibility for co-ordinating development of the new fighter aircraft, which the following month was named System 37, or *Flygplan 37*. In the fall of 1961, specifications for System 37 were approved by Stockholm's supreme commander. It was to be a totally new aircraft, designs based on the Draken airframe being discarded. It was to be innovative but not revolutionary – a very radical vertical take-off (VTOL) design with lift engines in the nose was shelved. It would be a multi-purpose machine with STOL performance, offering high flight-time availability and short turn-around time for repeated sorties. It *would* be a single-seater.

Col Lage Thunberg of the Air Staff is credited with

Overleaf:

Top left:
JA37 Viggen No.37360 of F17 wing at Ronneby in 1984.
Peter R. Foster

Bottom left:
Unlike the US Air Force, which camouflages its pilots' helmets, the flier at the controls of this AJ37 attack Viggen wears a pure white bone-dome. Here, forest green-brown camouflaged AJ37 is under tow during the annual Swedish Air Force open day at F16 wing, Uppsala, in August 1983. *Chris Pocock*

Top right:
A SH37 Viggen maritime surveillance aircraft of F13 wing at Norrkoping, on patrol. *Robert Lofberg*

Bottom right:
A gaggle of Viggens at F17 wing, Ronneby: from left to right, SH37, SF37 and JA37. The 'FC' code on the aircraft at the right is short for *Svenskt Provflygarcentrum*, the Flight Centre at Malmen.
Peter R. Foster

the formal decision on crew size. Although described by a critic as a traditional military man with no great belief in electronic devices, Thunberg was concerned with personnel costs from two-seat types like the J32 Lansen, the Buccaneer, and the Phantom. He concluded that a digital computer would do the job of a navigator.

A Concept and a Controversy

In November 1961, the Air Force issued directives for continued work on System 37. The following month, the long awaited decision on engine type was made.

Four different engine types had been discussed, all to be licence-manufactured in Sweden. These were the Rolls-Royce RB168 Medway, the Bristol Siddeley BO22R Olympus, the Pratt & Whitney JT4-29, and the Pratt & Whitney JT8D-22. It had always been a principle in Sweden not to manufacture (or licence) engines not chosen by any other customer and, for that reason, the short-lived Medway disappeared at an early stage. The JT4 was found not to be suitable for the new fighter. SAAB had been strongly inclined toward the Bristol Olympus, chosen in December 1959 by the RAF for its advanced TSR2 strike aircraft. Almost as if they forsaw cancellation of the promising TSR2, which came later in 1965, Air Force officers, especially Col Ake Suden, applied pressure which overcame the manufacturer's preference. In December

43

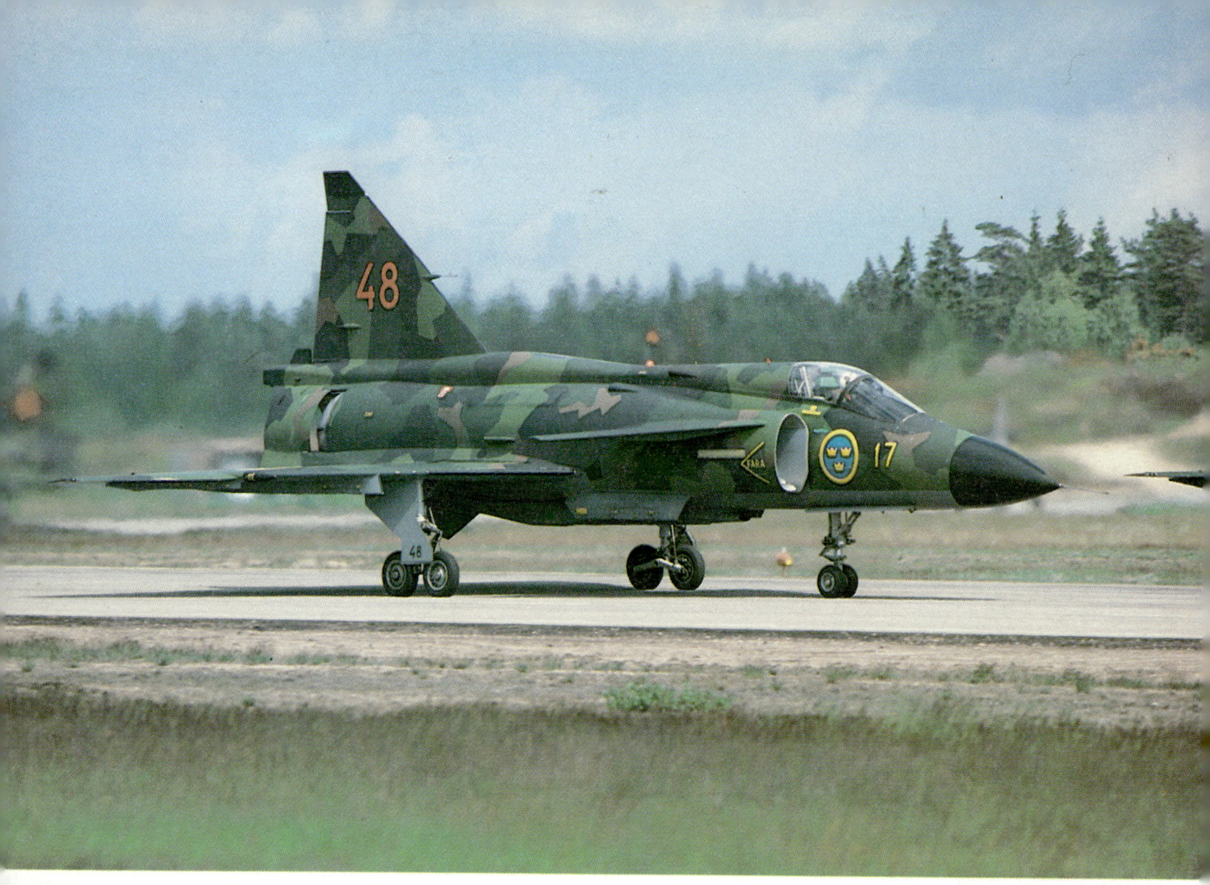
48
17
48

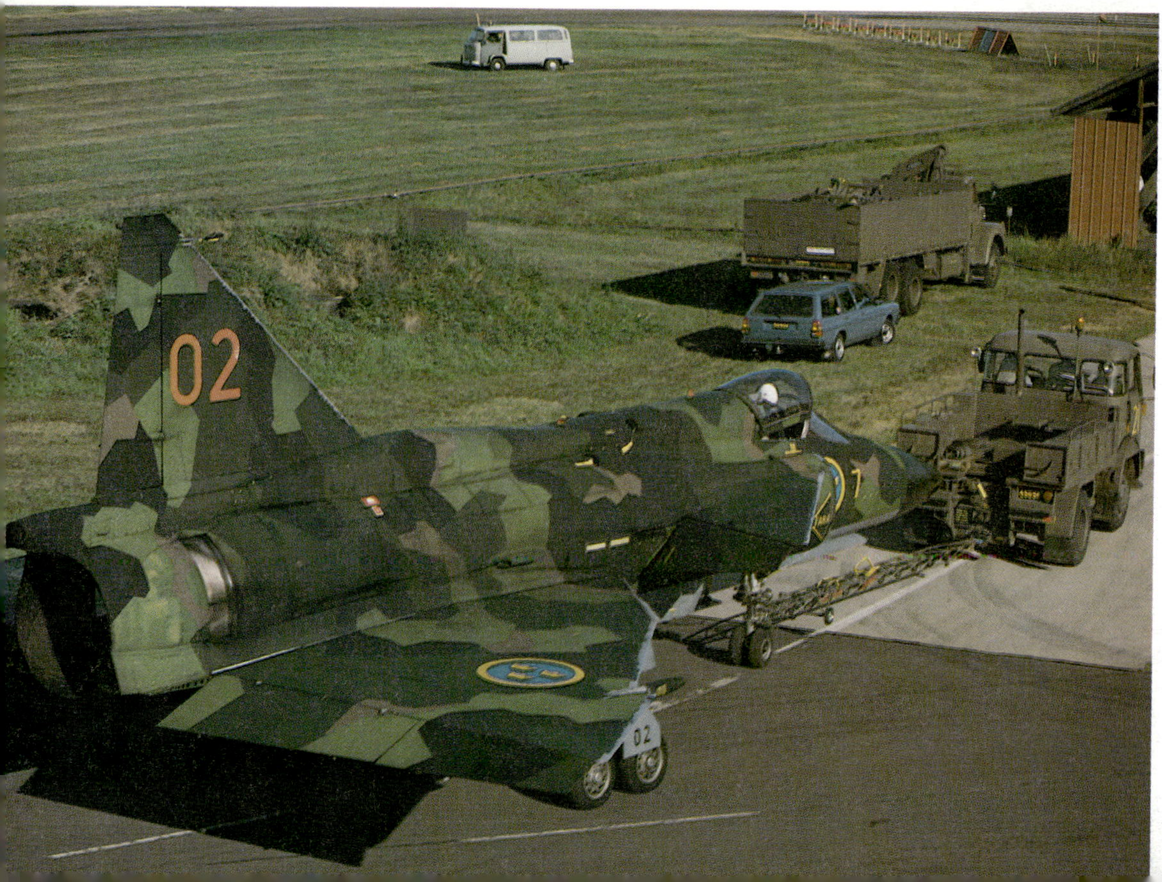
02
02

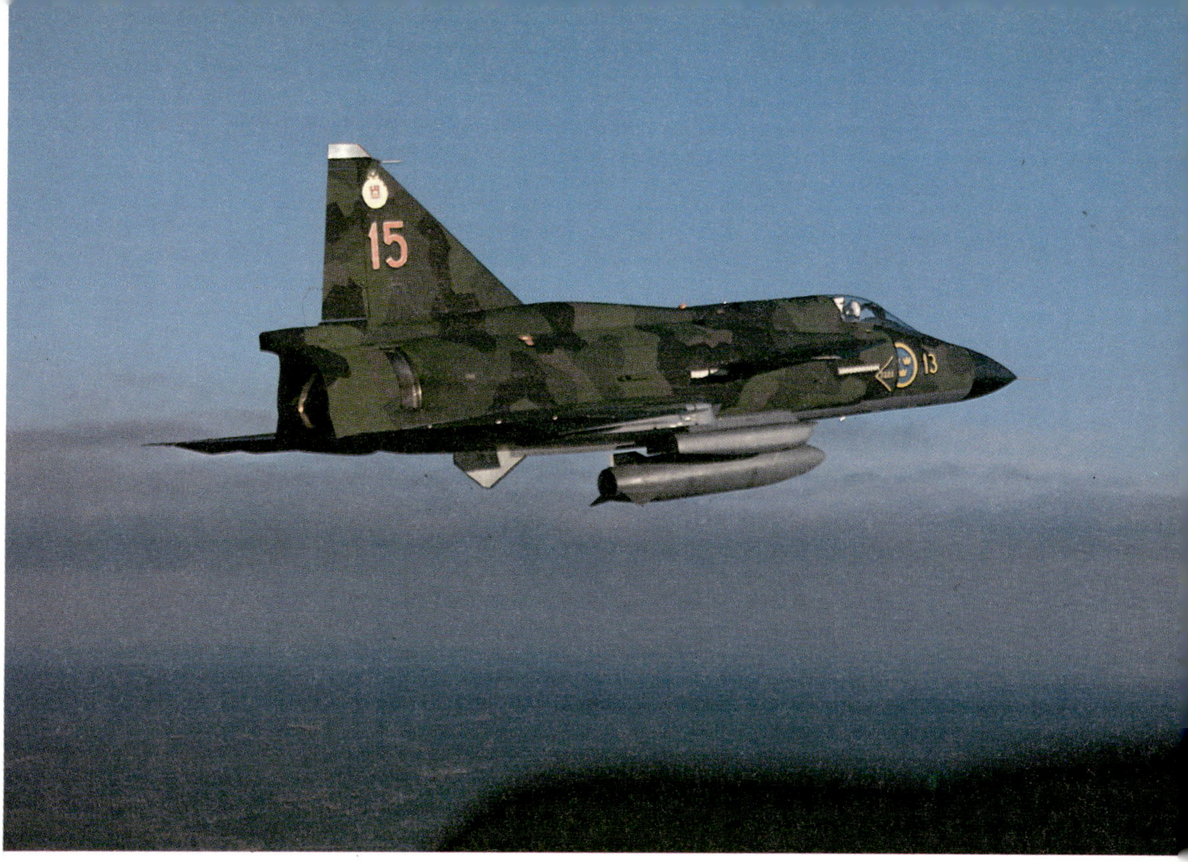
15
13

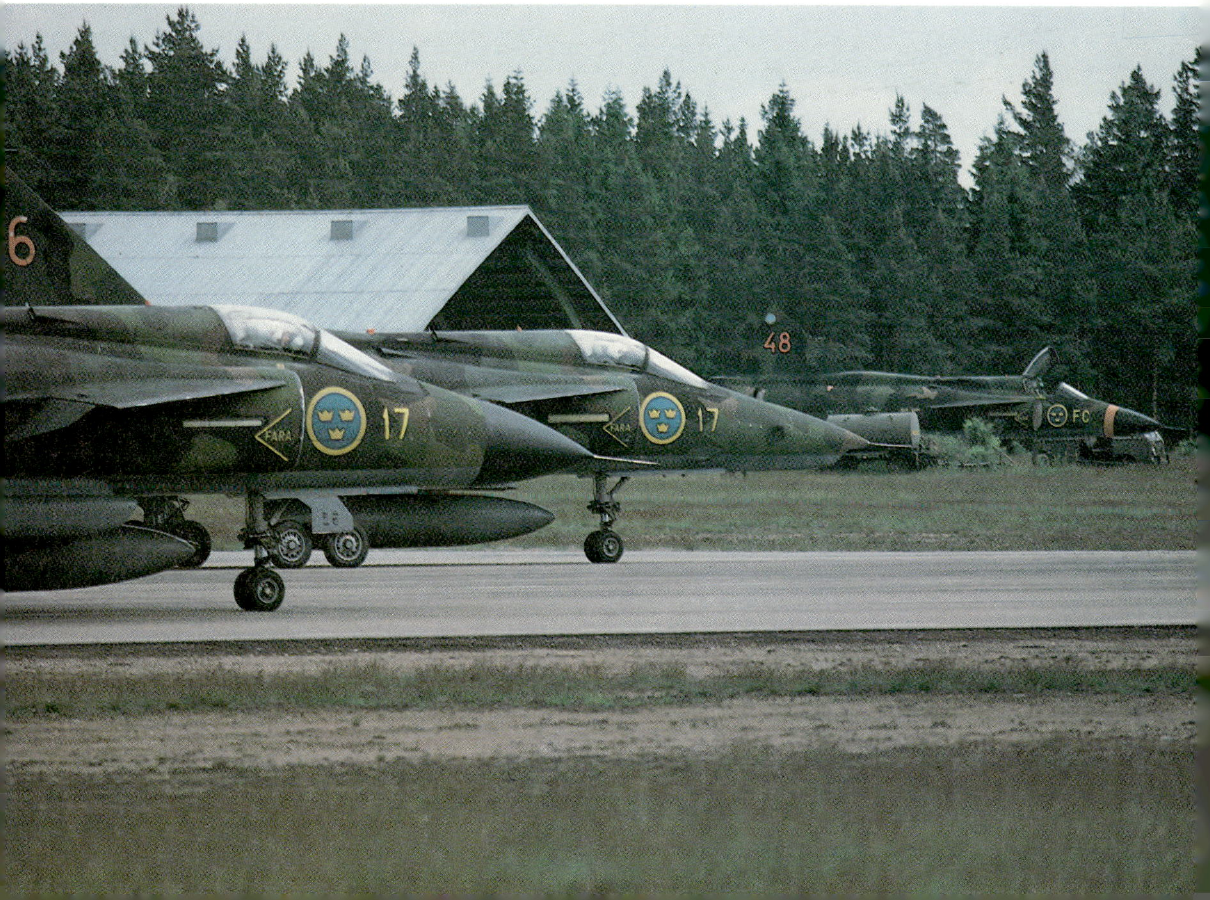
6
FARA
17
FARA
17
48
FC

1961 the Air Force determined that the plane would be powered by a military version of the Pratt & Whitney JT8D-22.

In February 1962, as pieces of paper began to flow out of Erik Bratt's shop, SAAB presented its final proposal for the fighter's aerodynamic configuration, using a canard surface in combination with a delta wing. It was also publicly announced at this point what had long been decided, what Colonel Thunberg had wanted, that the aircraft would be equipped with a digital computer. (No other modern aircraft except the Vought A-7 Corsair, almost contemporary with the Viggen but otherwise unremarkable as an advance in technology, is equipped with a computer that can assist the pilot in virtually every phase of flight operations.) In April, the obvious was formalised – that SAAB had been chosen as the prime contractor for the next-generation fighter. December 1962 saw the first detailed press release on what was now being called Project Viggen. Although the name is often translated as Thunderbolt, the word Viggen actually refers to a particular thunderclap made by the hammer of Thor.

From tinkering with ideas in the early 1950s, Erik Bratt's SAAB design team had moved to an airframe configuration which, by the early 1960s, was fixed in concrete, as one member of the design team puts it. Not that there was no flexibility, or that innovation came to a sudden halt. Indeed, in the same December 1962 wind tunnel tests which fixed the shape of air intakes, canard surface, wing and vertical fin planforms, a decision was made to shorten the Viggens fuselage by one metre (39in). But by 1963 the basic canard/delta planform was decided upon and men began to see what their airplane would look like when metal was cut. Lars Brising, SAAB's Vice President for Technology, and designer of the J32 Lansen, officially signed a company administrative memorandum in late 1962 authorising the Bratt team's concept to be built as actual hardware. By early 1963 factory tooling was progressing rapidly.

The Viggen programme, because of its enormous cost, had critics. The controversy seems to have come to a head in early 1964, when public and private debate was intense. (It was to recur during the 1965 elections.) On 1 April 1964 the sensationalist tabloid *Aftonbladet* published a front-page article, long in preparation, entitled 'Million Kronor Project A Fiasco?' Charges were made of poor planning and administration and it was asserted that Sweden was 'not rich enough' to afford a fleet of several hundred Viggens. (Contemporary plans called for delivery of 67 Viggens per year once production was underway; the Draken had been delivered at the rate of 45 airframes per year.) Cartoonist George Beverloo the following day published a caricature suggesting that the newspaper's publishers, not the makers of the Viggen, were misguided.

On 10 April the more prestigious newspaper *Dagens Nyheter* published an informed and thoughtful piece

Right:
Carrying dummy Rb.28 Falcon missiles, the first Viggen, 37-1, is seen at Linkoping in November 1968. Tandem, double-wheel main gear was to prove highly effective on the operational fighter. *Saab-Scania*

Centre right:
Carrying two Rb.04E air-to-surface missiles, the second development Viggen airframe, 37-2, flies through Nordic skies. *Swedish Air Force*

Bottom right:
37-2, the second pre-production Viggen, flying at low level. *Saab-Scania*

asking, 'Is Viggen Too Big A Project For Us?' Like the fits and starts of the early design period, debate over cost-versus-need would move in different directions while an actual, flyable airframe continued to taken shape. Some Viggen, in some form, was now inevitable but it was widely understood that only the initial production run of the AJ37 ground attack version was 'safe'. Political debate might yet kill the planned follow-on JA37 interceptor.

Once a mock-up of the Viggen was completed and shown, a more serious, less public debate ensued over whether maintenance of such a sophisticated design would be practical. To be sure, the Viggen was – and is – sophisticated. But to put developments elsewhere into context, although no one knew it yet, on 22 February 1964 (while Viggen existed only as a full-scale mock-up) the Lockheed A11 – progenitor of the hyper-advanced SR-71 reconnaissance craft – made its first flight at a desert airstrip in Nevada.

'We knew there would be doubters', says Olof Esping, an engineer among the SAAB force committed to Viggen (which rose from 300 workers to 2,000 between 1963 and 1965). 'Plenty of people were saying, let's just buy Phantoms instead. But we knew that our aircraft would be a winner.'

The mock-up retained a ventral fin which, in wind tunnel tests, had seemed necessary for stability, particularly in tight turns. Debate would continue but the prototype now moved relentlessly toward roll-out.

The First Viggen

The prototype Viggen bore the serial number 37-1. Seven further development airframes followed before the first production AJ37 was to fly with the traditional five-digit *Flygvapen* serial (or FV number) 37001.

37-1, which had no radar, was all silver, with the triple crown national insignia on its nose. It was equipped with a long pitot probe not found on production machines. Also called 'SAAB 37-1', it was rolled out with appropriate ceremony at the company's Linkoping plant on 24 November 1966. Engine trials were begun on 16 January 1967, and the first taxi tests for 37-1 followed on 31 January 1967.

Few men enjoy the mixture of talent and luck to

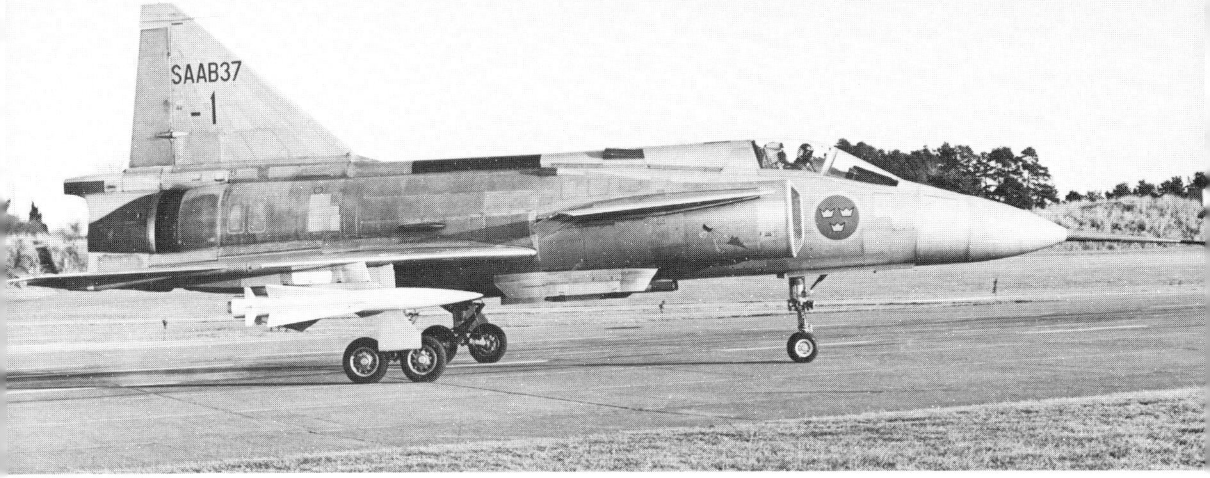
SAAB37
-1

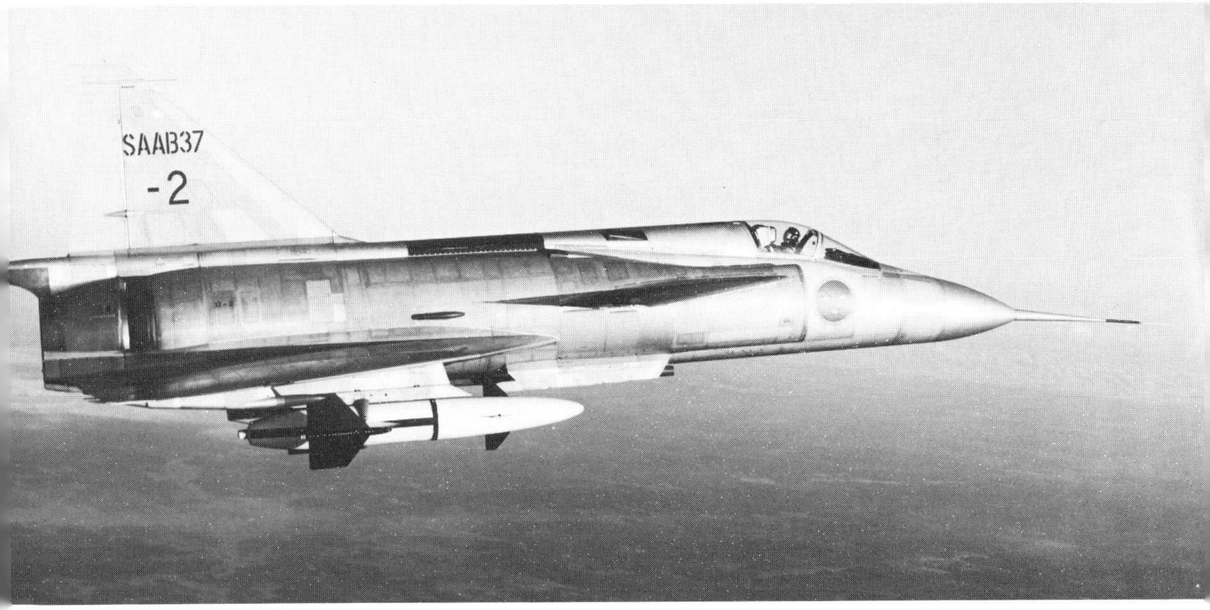
SAAB37
-2

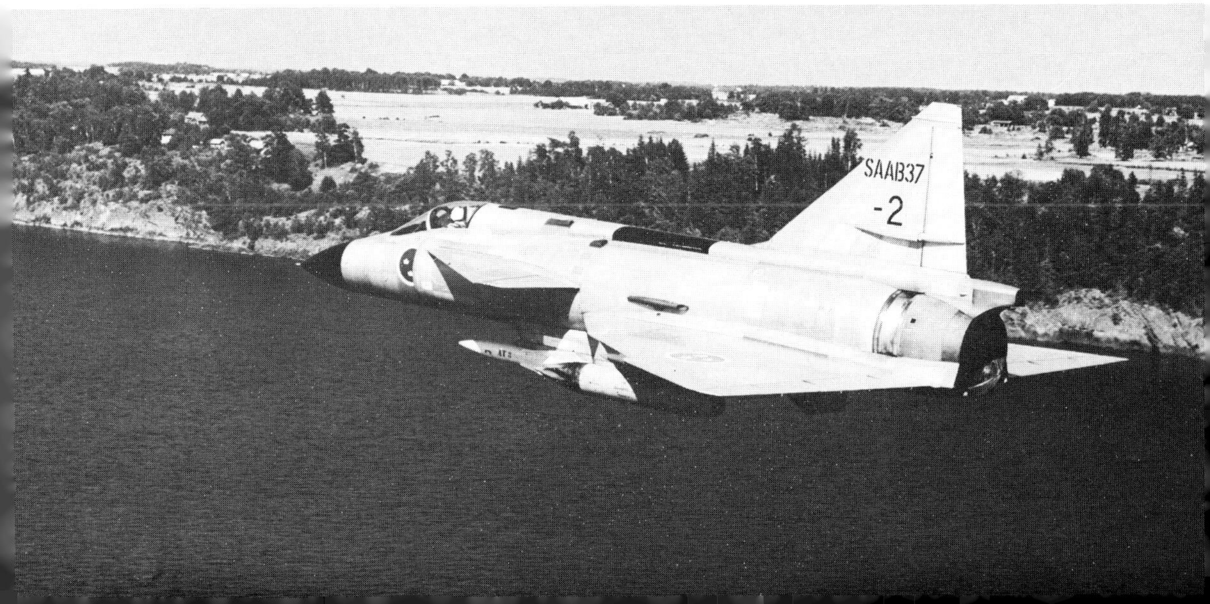
SAAB37
-2

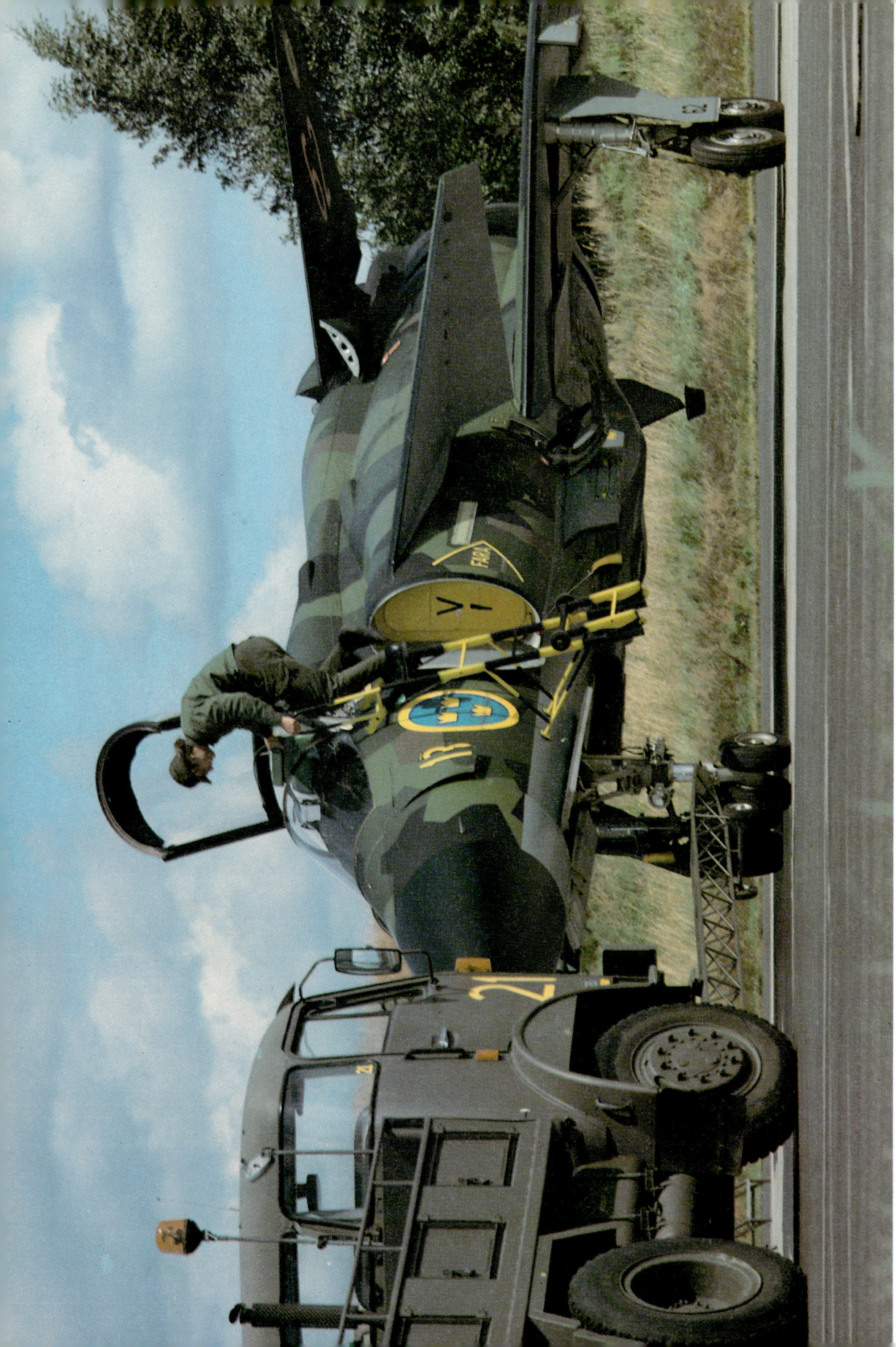

make history's first flight in a world-class fighter aircraft. Wing Cdr Roland P. (Bee) Beamont, who took English Electric P1 Lightning WP860 aloft on 4 August 1954, is a renowned 'total aviation person'. Robert C. Little, who piloted McDonnell F4H-1 Phantom No 142259 on its maiden voyage on 27 May 1958, and who failed to exceed the speed of sound then because of a technical hitch has gone on to become a company vice president. But how many remember Paul Millett and Nils Meister who, on 14 August 1974, first flew the Panavia Tornado? Into the tight hallowed company of 'first flight' veterans, to take Viggen aloft for its initial trip, stepped gaunt-faced, 45-year-old SAAB test pilot Capt Erik Dahlstrom.

Born on 27 July 1922, married, living with his wife and two daughters in Linkoping, Dahlstrom seemed eminently the right man with the right stuff. He had been flying for 25 years. He'd begun his career in 1942 with F5 wing at Ljungbyhed and had logged time in no fewer than 42 different aircraft types. He had begun his career with SAAB as commander (the military term is used) of its Flight Test Department in 1959. On 14 January 1960, in a J35D Draken, he had become the first Swede to exceed Mach 2.

On 8 February 1967 Dahlstrom took Viggen 37-1 aloft for an unremarkable first flight. Only company personnel and a handful of selected onlookers were present. Dahlstrom was under instructions not to fly supersonic on the first flight, and did not. He made a pass over the field which impressed his audience and gave them first exposure to the JT8D engine's throaty growl. No major technical problems were found and Dahlstrom reported that the aircraft handled well. For various reasons, including the fact that Viggen programme funding was still a touchy issue in Parliament, Dahlstrom did not fly 37-1 for the benefit of the press until 6 April, when the flight test programme was in full swing.

The Viggen took some getting used to. Its high flare angle on touchdown is far from unforgiving but, without experience, it invites the pilot to over-react and brush the lower rear tailpipe against concrete, or worse. An audio warning of what would be a high-angle stall in most aircraft, but is not in the Viggen, was fitted after Dahlstrom brushed the lower rear fuselage against the runway.

The second Viggen, 37-2, flew for the first time on 21 September 1967. This machine was to be used for trials of the Viggen's integrated flight control system, and it was to enjoy a long and full life. It was later renumbered 37-21 and first flew with that number on 4 June 1974 as the control system test aircraft for the second-generation JA37 interceptor. By 1981, long

Left:
An AJ37 Viggen strike aircraft in mottled green-brown forest camouflage being readied for a mission.
Swedish Air Force

Below:
This rocket projectile pod hanging from the wing of second pre-production Viggen, 37-2, typifies various loads which were envisioned for the aircraft early in its career.
Saab-Scania

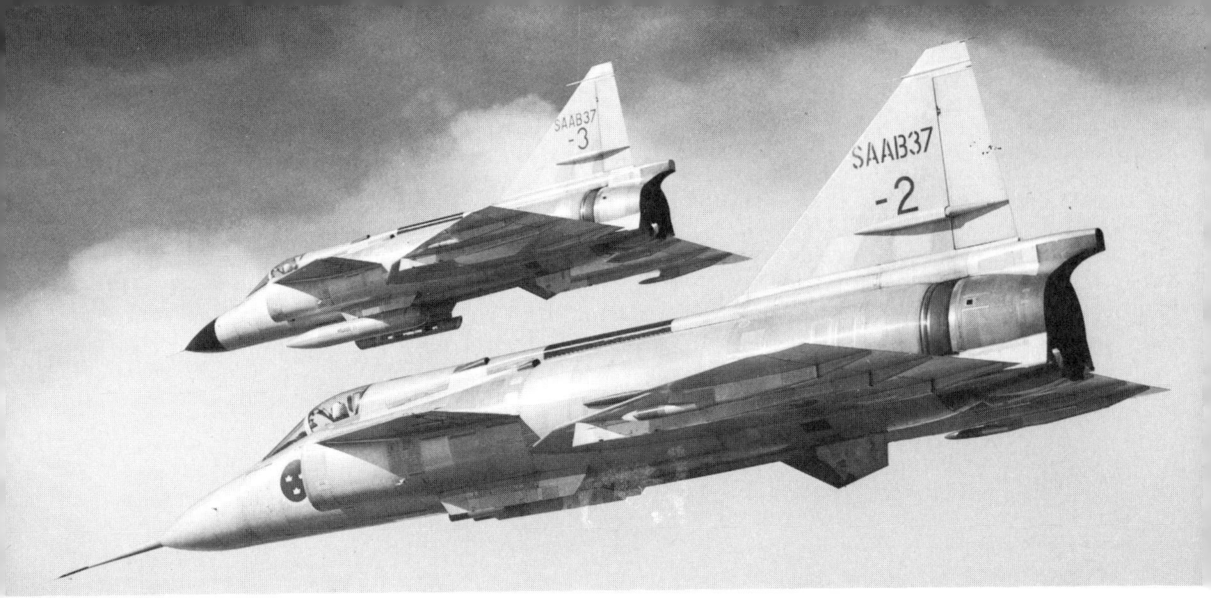
SAAB37
-3
SAAB37
-2

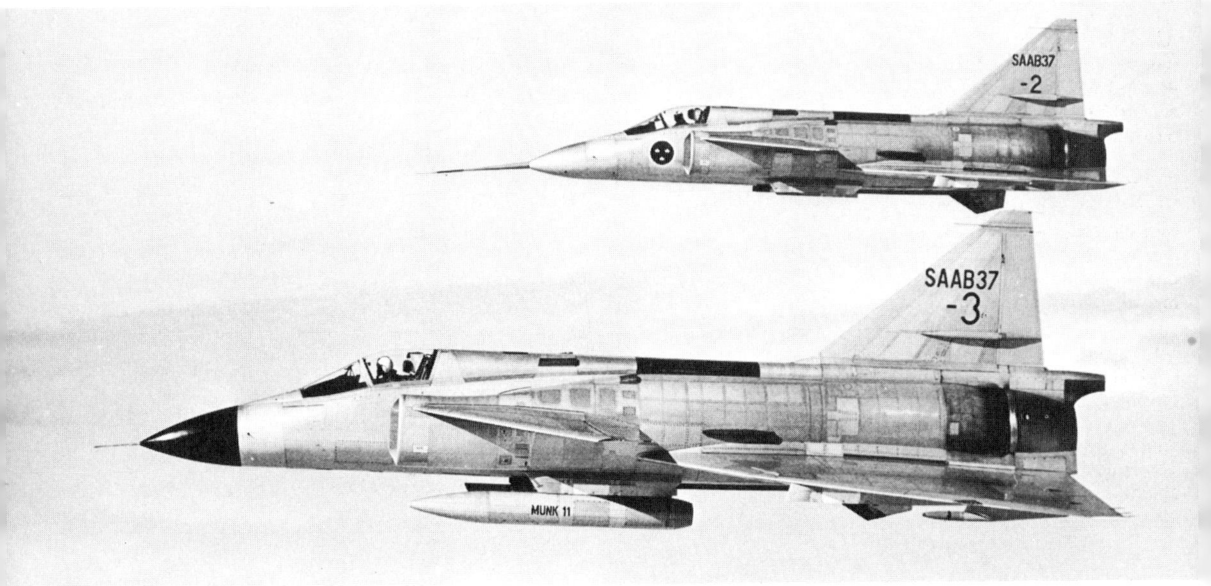
SAAB37
-2
SAAB37
-3
MUNK 11

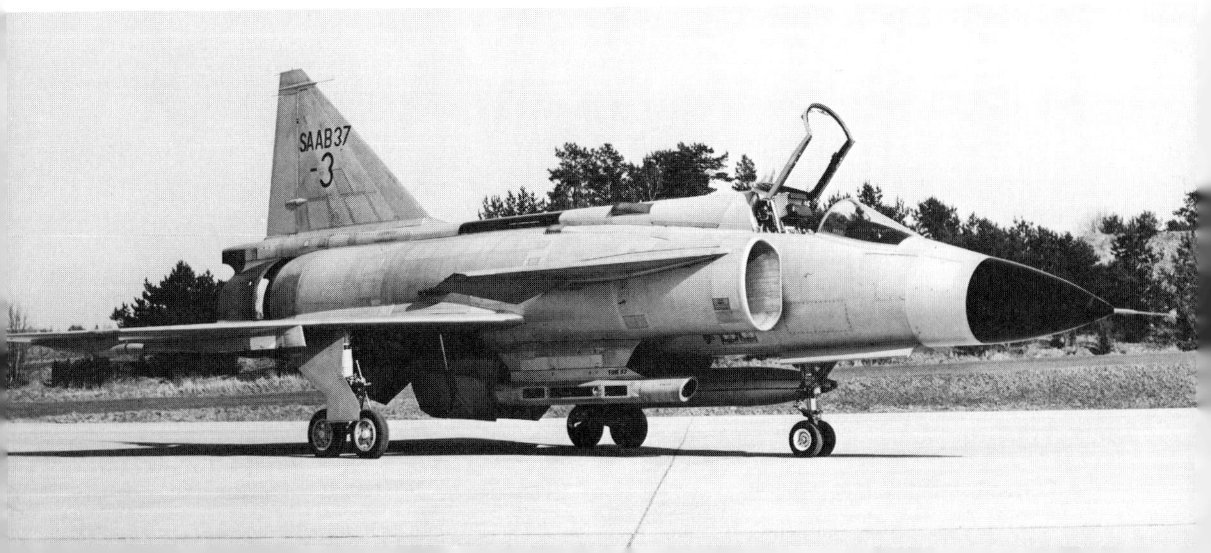
SAAB37
-3

Above left:
As an ambitious flight test programme gathered momentum in 1967-68, Viggen airframes began to proliferate. Aircraft 37-2 and 37-3 (in background, carrying anti-collision pod) here fly formation together. *Saab-Scania*

Left:
Viggen 37-3, in the foreground carrying a collision-avoidance pod, was the first equipped with a full avionics system. Its nose radome shape is markedly different from 37-2 in the background and is as adopted for the operational AJ37. *Saab-Scania*

Below left:
Canopy open, developmental aircraft 37-3 is ready for continuing flight tests at Linkoping. *Saab-Scania*

Above:
The fourth Viggen built, 37-4, was the first to incorporate all of the wing planform features of the production aircraft. *Saab-Scania*

after Viggen was firmly settled in its operational life, this high-hour airframe evaluated the *elektroniskt styrsystem* (EES), or fly-by-wire system, for the JAS39 Gripen.

Accelerated Testing

By early 1968, as its stable of test ships grew, SAAB began carrying out up to 35 test flights a day, aircraft at times making five sorties a day. Seven pilots, five from SAAB and two from the Air Force, were now committed to the programme. Debates over funding and other issues continued, and while it was clear that the AJ37 ground attack machine would enter service – by 1967, said the Air Force; by 1971, said Parliament – the future of reconnaissance and surveillance variants, and of the later-stage JA37 interceptor, seemed clouded. Linda Korinek, a noted economist, asserted that 'Viggen will drown us' (in deficit spending). Many such claims were exaggerated, Brising often reminding listeners that SAAB was developing the aircraft at predicted costs, but the maker and the Air Force were anxious to carry out a development programme to be proud of. Tests in the low altitude portion of the flight envelope confirmed that the unorthodox design was relatively insensitive to gusting, moderately easy to handle, and surprisingly stable. With experience the aircraft was found to be easy to hold on glide slope under difficult crosswind and power-adjustment conditions, and even a novice could land successfully at a 16.5° angle at 105mph (172km/hr). Whatever it cost, Viggen worked.

The number three airframe, 37-3, was the first Viggen equipped with a full avionics system, which costed roughly $400,000 (£140,000). No 37-3 also flew with a collision-avoidance pod, invaluable in a test programme but not scheduled for the operational AJ37. First flown on 29 March 1968, 37-3 too would have long life, eventually being rebuilt and renumbered 37-31 to serve as the radar test vehicle for the second-phase JA37 interceptor. In 1968 though, that second step in Viggen's life still required further debate. Machine 37-4, the fourth in the Viggen stable, may have fuelled debate. Intended to evaluate powerplant, air intakes and tail cone, it crashed shortly after its 1968 maiden flight and had to be written off. Viggen 37-5 followed and was used to test stores separation and jettisoning of drop tanks.

Thrust reversers, so essential to the required STOL performance, also produced an odd side effect. Though there seemed to be no purpose to it, Viggen could be taxied backwards! This would turn out to be an unplanned benefit: in this manner, Viggen could more easily be manoeuvred in and out of the narrow hardstands that line the runways on Sweden's road bases.

Aircraft 37-6 was assigned to overall systems tests

with electronics and weapons. This machine was later rebuilt and renumbered 37-61 for further weapons evaluation. Viggen 37-7, seventh and last in SAAB's early test stable, would later be rebuilt and renumbered 37-71 as the prototype Sk.37 two-seat trainer. Scheduled later for operational serial number 37800, the first in the service Sk.37 series, it crashed and was written off before the five-digit *Flygvapen* serial could be painted on.

On 5 April 1968 the Swedish government confirmed its decision to order 175 aircraft of the AJ37 strike fighter variant. One hundred aircraft had been decided upon in principle on 21 March 1967, and a further 75 resulted from a further year's planning and negotiations. The airframe contract worth 1.69 billion kronor (£136 million) was the largest single order ever placed with Swedish industry.

As flight testing and operational evaluation continued through 1968-70, Viggen was now a fait accompli to its detractors and was rapidly proving a joy to its strongest advocates. Each new step in the flight development programme helped to fix the design of the future operational fighter. For example, the fourth development airframe, 37-4, was the first to have an extended wing leading edge outboard of the wing fence. This benefitted airflow over the ailerons and became the standard planform, giving the wing its now-familiar 'sawtooth' leading edge shape. Tests progressed well and plans were finalised from the AJ37 ground attack machine to enter service with F7 wing at Satenas in 1971. SAAB and its subcontractors had gone from engineers' daydreams to successful flight of a fighter aircraft of enormous technical prowess.

Above left:
SAAB's test stable at Linkoping facility in April 1969. In the foreground is the sixth pre-production Viggen, 37-6. *Saab-Scania*

Left:
One airplane is missing. Five of SAAB's six test aircraft, including 37-6 (foreground) in April 1969. *Saab-Scania*

Above:
The whole purpose of developing and test-flying the Viggen was to get the operational AJ37 into service, and the first wing to receive production AJ37s was F7 at Satenas. These are a quartet of F7's early operational AJ37s in flight in 1973. *Saab, via Jerry Scutts*

5 Technical Description

Above:
The Viggen is distinguished from all other combat aircraft by its canard double-delta configuration, here standing out against Swedish forests as two AJ37 ground attack machines from F7 wing at Satenas spring out at low level in 1973. *Saab, via Jerry Scutts*

A 'family day' at a Viggen fighter base usually is open only to the close relatives of those who service and fly these aircraft, not to the general public. Since wives and children have more than a little familiarity with the Viggen, it might be supposed that a kind of 'ho humm, nothing special' attitude might be theirs when looking close-up at the aircraft. It is not so. *Nobody* takes the Viggen for granted. 'It is an impressive sight every time', says Lena Lindstrom, wife of an Air Force officer.

Even so, if the Viggen is to be understood, awe and wonder must give way to dry technical jargon.

What follows is a brief description of some of the key aspects of a unique aircraft which resolves an apparent contradiction – a multi-role system at low purchase and operating costs. Because so much was expected from a single airframe design and because it is part of SAAB's tradition to be innovative, it will not be surprising that some of the Viggen's features are remarkably unorthodox.

Basic Flying Platform

The basic Viggen design was intended to provide high speed and good performance together with the all-important short take-off and landing (STOL) capability wanted by *Flygvapnet*. The engine was chosen to give low fuel consumption for long range at moderate speed and very high thrust with afterburner. The nerve-centre of the basic airframe design is the digital computer, still among the most advanced found on combat aircraft, the program of which can be changed to suit the specific demands of the various aircraft versions. It was seen from the start that the

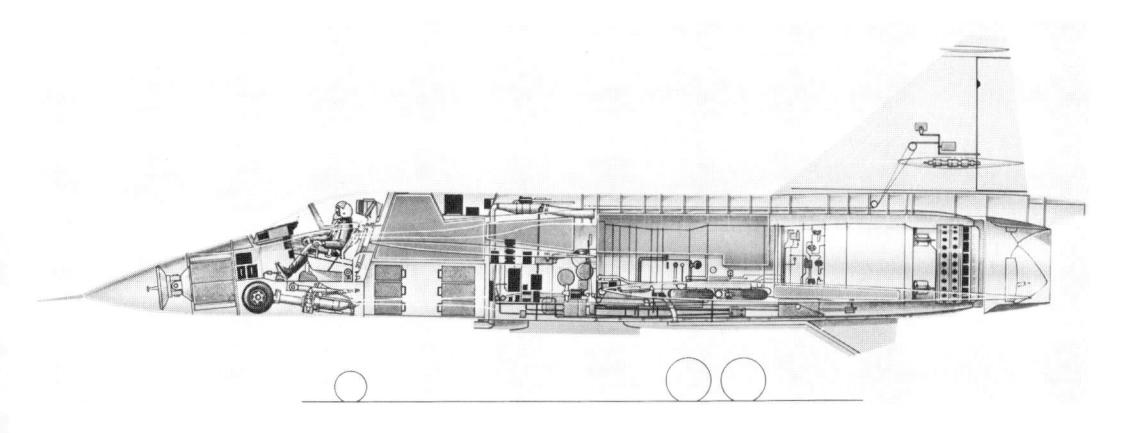

Above:
The basic Viggen design is compact and sensible. This cutaway drawing of an AJ37 Viggen shows how effective use was made of internal space, from the forward-folding nosewheel beneath the pilot to the afterburner-equipped tailpipe exhaust. *Saab-Scania*

Viggen would be developed in attack, interceptor, reconnaissance and trainer versions.

The Viggen is the first genuine canard military airplane to enter production for more than half a century. The delta-wing canard shape sometimes causes the Viggen to be called a 'biplane' due to the lift qualities of the forward surface, or foreplane. It also assures the Viggen's high performance and provides high lift for the need STOL capabilities.

The basic airframe, designed to withstand load factors of up to 12g, makes extensive use of honeycomb panels and metal bonding. Surfaces of the very intricate foreplane and wing conceal widespread honeycomb panels of various sizes used throughout, except on the leading edges. The result is an airframe design far lighter than it appears, honeycombing being combined with extensive use of titanium and titanium bolts to save weight.

More than a hundred inspection hatches and covers on surface locations of the Viggen airplane provide ground crews with remarkably easy access to internal systems. The nose cone has a quick-release handle and may be pulled forward on tracks for easy inspection of the radar. The rear fuselage section, with thrust reverser and ejector air intake, may be removed to permit quick engine change, a total of only six man-hours being required. The wing, fin and foreplane surface attachment points are few and the time required for a change is therefore short. Since the Viggen sits rather high, the vertical fin has been made foldable in order to permit passage through low hangar gates. The one-piece cockpit windscreen is designed to withstand the impact of a bird strike from a bird weighing one kilogram at 1,000km/hr.

Landing Gear

The Viggen's landing gear has been designed for a high rate of sink, as much as 5m/sec, to make carrier-type, no-flare landings possible. It is made of high strength steel.

The main gear is mounted in the wing and retracts into the wing and fuselage. A tandem wheel configuration was used to reduce the depth of the wheel well, and the height of the gear is shortened during retraction by compressing the shock absorber.

The tandem wheel arrangement has other advantages. A softer shock absorption is obtained and reduced tyre pressure may be used, the latter being important in connection with 'road base' operations.

Radar

The PS-37/A radar system used by the AJ37 attack Viggen was developed by L. M. Ericsson Company, which also built the radar for the Lansen and Draken. This multi-purpose monopulse radar is designed in solid state (except for microwave tubes) and is used for ground mapping, air-to-ground ranging, obstacle warning and air-to-air interception. The terrain-following capability, which requires addition of a small terrain-following computer, seems to be a retrofit which enhances the radar's usefulness.

The radar scope for the pilot's use, manufactured separately by Svenska Radio, is located beneath the head-up display in the middle of the instrument panel. For ease of maintenance the radar, which has more than 3,000 electronic components, is subdivided into 13 replaceable units, each in turn composed of six to eight sub-units.

Theoretical design on the radar system began as early as 1958, and ground tests began in 1961. A prototype PS-37/A system was completed in 1965. With characteristic thoroughness, several Lansen aircraft modified by a team under SAAB's Gosta Niss were used in actual flying tests of the Viggen radar in 1966, before any Viggen had flown. The first such Lansen, nicknamed '32 Alpha' (FV serial number 32080) was followed by '32 Gamma' (FV serial 32280) which had the Viggen nose radome. The radar manufacturer's literature contains a cryptic notation that 'verifying air tests of the PS-37/A radar were made in a (Viggen)' in 1968 and 'series production of the radar began in 1970'.

The PS-37/A radar is constructed for reliability and for such good performance that navigation towards, discovery and identification of, and attack against the target can be carried out with great precision and without loss of time or operative performance, even during powerful vibrations within a large frequency band and in spite of rapid changes within the temperature range of $-40°C$ to $+60°C$. The PS-37/A can detect targets on land, sea and air at great but undisclosed distance and can separate these from their surroundings and from each other. The system's long range is achieved by, among other things, a high power output, an antenna of 'large' diameter (apparently about 32in), and specially-shaped feed which gives an antenna lobe with high grain and small side lobes. The radar has high resolution and what Ericsson calls 'a new interference protection technique' (not further divulged) which resists electronic countermeasures.

For the JA37 interceptor variant, L. M. Ericsson's MI Division at Moelndal developed the PS-46/A X-band pulse doppler radar, described by SAAB test pilot Gost Sjostrom as 'a quantum improvement'. Says Sjostrom, 'all the buttons needed for intercept are readily reachable and recognisable by feel!'

With the potential to have ground mapping and sea surveillance modes added – the extent to which either is now fitted on the aircraft has not been disclosed – the PS-46/A modular software controlled radar system provides four air-to-air modes: target search, track-while-scan, continuous track, and target illumination, using an X-band CW illuminator also built by Ericsson and integrated with the radar set. The illuminator is used with the British Aerospace Sky Flash air-to-air missile.

The Viggen interceptor radar is made up of 10

Below:
Nicknamed '32 Gamma', this SAAB J32B Lansen fighter was extensively modified by a team headed by the company's Gota Ness as an in-flight testbed for the PS-37/A radar system developed for the Viggen. The radar installation gave this Lansen's nose a distinctive downward-drooping appearance.
Saab, via Robert Lofberg

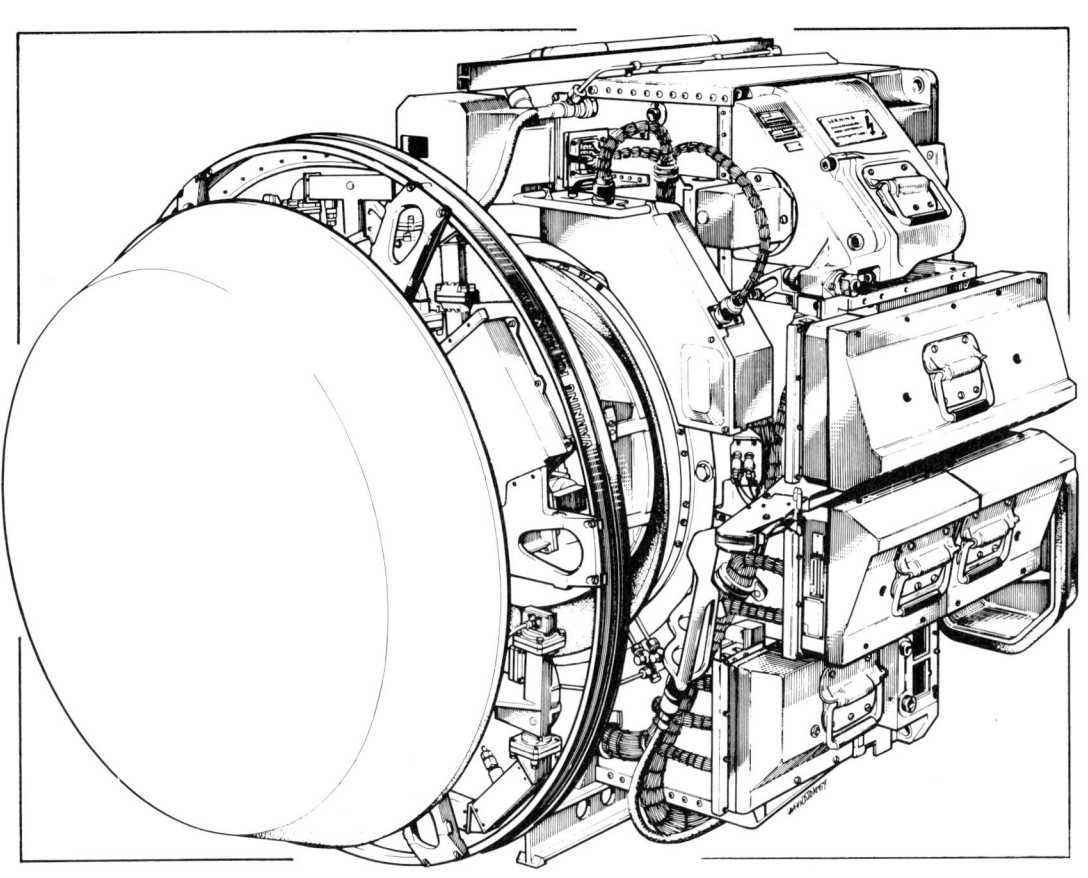

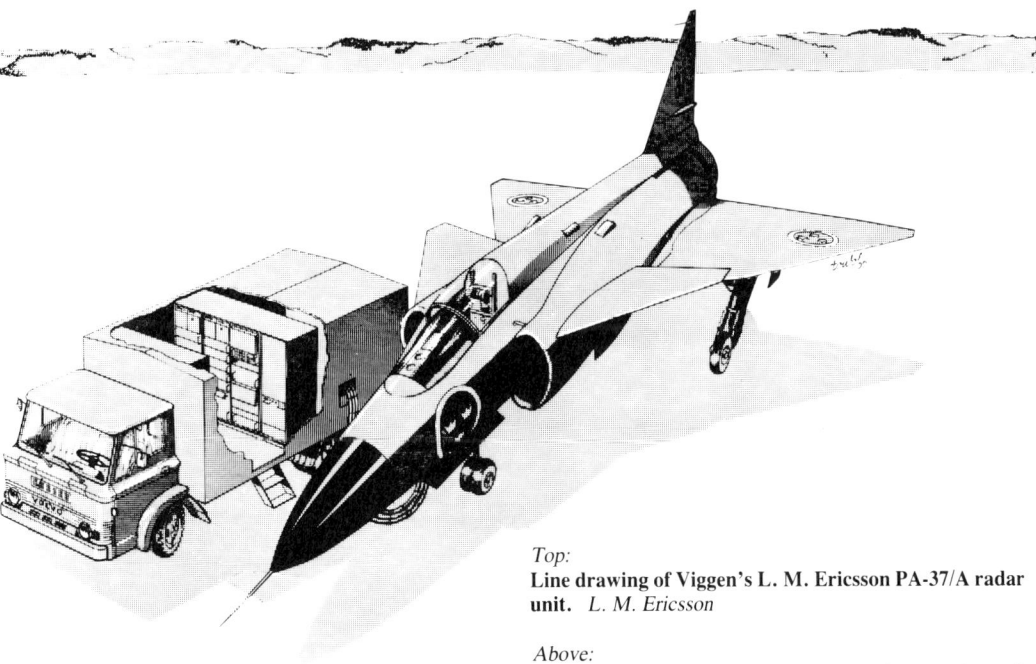

Top:
Line drawing of Viggen's L. M. Ericsson PA-37/A radar unit. *L. M. Ericsson*

Above:
Drawing of Viggen with test equipment for its radar unit mounted in a vehicle beside the aircraft. *L. M. Ericsson*

Below:
The Viggen nose cone slides forward for access to the PS-37/A radar unit. *L. M. Ericsson*

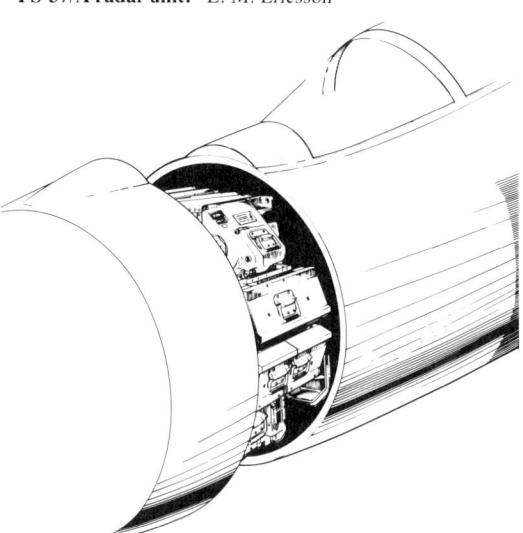

line-replaceable units supported by a lightweight frame. The entire radar is attached to the aircraft bulkhead with four bolts and may be replaced quickly as a unit. The 660lb system uses a 28in antenna with very low sidelobes, sidelobe blanking and gap filter and IFF antennas integrated. The radar has a detection range of more than 31 miles in a look-down mode against typical military aircraft targets.

Below:
Considerable work went into the ergonometric cockpit design of the Viggen. The cockpit of the instructor's position in the rear of the two-seat Sk.37 Viggen trainer is simple and efficient, but does not have either weapon controls or navigational aids, including moving maps, found in the cockpit of the single-seat AJ37 attack aircraft. *Saab-Scania*

Right:
The second prototype of the JA37 interceptor aircraft displays the 30mm Oerlikon KCA cannon which is mounted snugly beneath the fuselage of the aircraft. *Saab, via Bill Gunston*

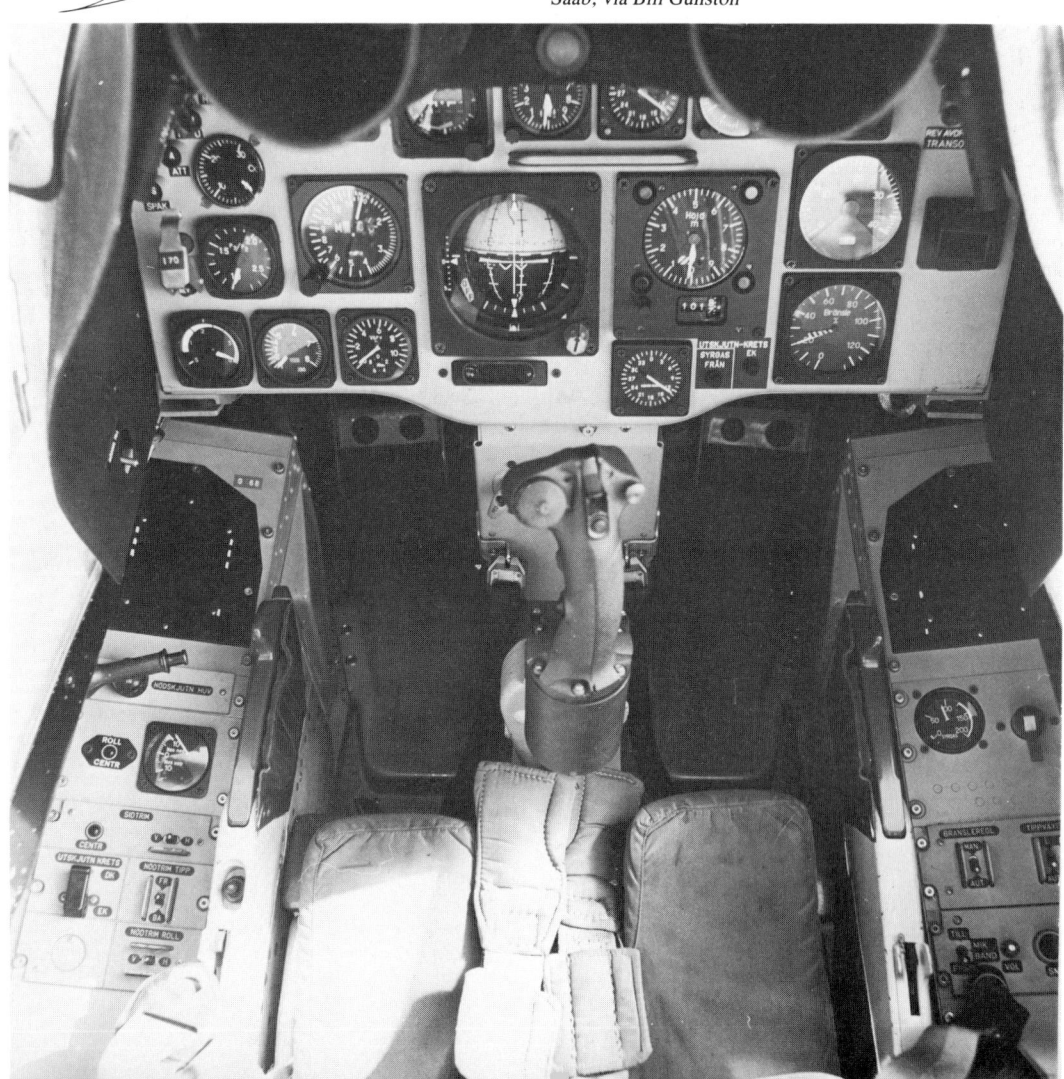

Cockpit

The Viggen's ergonometrically designed cockpit is the result of a conscious research and development programme to reduce pilot workload, eliminate unnecessary pilot strain and permit absolute mastery over controls and instruments. The cockpit is logically and neatly arranged; the number of instruments has been kept to a minimum and the instrument panel is close to the pilot for easy readout. Few fighter pilots sit as comfortably or enjoy as much room as the man who flies the Viggen. Even wearing his full array of flying gear, the pilot enjoys almost a luxury of space, expecially leg room where the large flat rudder pedals form a V at the base of the instruments.

Great care has been taken to relieve the pilot's burden during each phase of a mission. For example, there is the stress placed on the importance of the automatic navigation system from which all steering and navigation commands are presented to the pilot on a Head-Up Display (HUD). The HUD, developed by Svenska Radio, is located in the pilot's line of sight. It has two main tasks: to present information for navigation and landing, and to provide an electronic sight during approach and attack in those combat situations where an optical sight is needed. The picture contains integrated information as to attitude, course, height, range, etc. An optical system and a semi-transparent mirror are used to project the picture into the pilot's line of sight, so that it is seen against the surrounding background. The picture is focused on infinity, thus standing out sharply against the background and avoiding eye fatigue. In order to maintain an acceptable contrast between pictures and background even when the illumination of the latter varies from very low to extremely high values, a special cathode ray tube (CRT) in conjunction with a photo cell is used. The maximum brightness of this tube is very high and is controlled by the current from the photocell which measures the background illumination.

Of course, the pilot also has a Head Down Display, located beneath the HUD. It has two different functions, which can be selected by the pilot. It can either present flying data or be used together with the radar as auxiliary navigation equipment and electronic sight. In the latter case, it shows a radar map of the terrain in front of the aircraft. On this picture there are super-imposed symbols which help the pilot to interpret it. Another symbol shows where the target will appear. When target contact has been established further symbols enable maximum use of the available weapons. Flying data is also superimposed so that the pilot need watch only one display. The Head Down Display is so bright that no shielding is needed.

One problem: radar information requires a relatively long persistence in order to yield a complete map. On the other hand, the symbols must have as short a persistence as possible, lest they mask the radar picture as they move across it. This problem was solved by using a cathode ray tube (CRT) with the unique property of showing two pictures with different persistences. One is practically zero and the other can be varied so that very long persistence can be achieved.

The cockpit air conditioning system is designed to maintain a desired pressure level in the cockpit and the

Above:
Test-firing Viggen's cannon from a specially-built centrifuge at the Vidsel missile test range. The 30mm Oerlikon KCA cannon has twice the effective range and eight times the impact energy of earlier 30mm guns. *Saab, via Mike France*

electronic compartments and to create a suitable temperature and humidity level for pilot and equipment. High pressure bleed from the engine is passed to the system through heat exchangers and cooling turbines for proper pressure and temperature reduction. Thereafter the humidity is reduced in a water separator and the air is distributed to the various compartments. A special connection is available for ventilation of the pilot's suit. Desired temperature may be selected by means of a thermostat control. For high alert conditions, when the engine is not running, the system is supplied with low pressure air from a ground unit.

Below:
A dramatic take-off shot of an AJ37 Viggen, in this instance piloted by Captain Anders Lefvert of F7 wing at Satenas, demonstrates the performance characteristics which made a zero-zero ejection seat for the Viggen essential. *Saab-Scania*

Gun

The second-generation Viggen, the JA37 interceptor, is armed with a gas-operated Oerlikon KCA 30mm automatic cannon which fires 22 rounds/sec and which fits snugly in a low-drag fairing beneath the belly of the aircraft. The cannon resulted from a long and costly development programme and from exhaustive tests, including firing tests from a specially-built centrifuge at the Vidsel test range. Claimed to have twice the effective range and eight times the impact energy of previous 30mm cannons – exact figures are not discussed – the weapon is designed to require minimal maintenance. On a typical intercept mission, the JA37 will carry two British Aerospace Sky Flash semi-active radar homing (SARH) air-to-air missiles, two AIM-9L Sidewinder infra-red (IR) heat-seeking air-to-air missiles, and the cannon.

Some Swedish pilots have expressed initial disappointment at the cannon's low rate of fire, which

Above:
Brightly painted and extensively modified, this SAAB J32B Lansen fighter (FV serial number 32502, also known as 37-8) was flown by pilot Milton Moberg to test the pilot ejection seat designed for the Viggen. No 32502 actually became a testbed for a variety of pilot escape systems developed by SAAB before the Viggen's system was finalised. *Saab, via Robert Lofberg*

is almost the same as the Aden, though the Oerlikon is a heavier gun, firing heavier shells. The gun fires in a remarkably straight trajectory, forming a compact 'spread' of fire. Its long range is important, especially if the aircraft being engaged by the Viggen has a gun in its tail. Also important is the cannon's potential for use in head-on combat situations, directed by the aircraft radar.

Escape System

Developed by SAAB based on experience with the Draken, the pilot's escape system in the Viggen is a rocket-powered ejection seat able to save life anywhere from zero-speed/zero-level to high speed and altitude. Ejection is initiated by the pilot pulling levers on the side of the seat.

In a society which places crucial value on human life, and one where the estimated £300,000 cost of training a Viggen pilot cannot be afforded more often than necessary, it was seen from the beginning as imperative that the Viggen should have an effective system to save its pilot in a mid-air emergency. Since a pilot could be injured or disabled, the system would have to function without his help once he had taken the first step of yanking the levers. Thus the rest of the escape system is automatic. All electrical circuits and some mechanical features of the escape system are purposely redundant for the sake of safety. Manual bypass of critical functions is a separate option.

The system was designed so that the Viggen's canopy would automatically be tossed away at the beginning of the ejection sequence. Both the excape system, and the general layout of the Viggen's cabin design, however, were deliberately arranged so that the pilot could 'punch out' through the canopy if necessary. (The author recalls sitting in an AJ37 and observing that his kneecaps would easily make it, if it had to be done that way. Pilots are known to be mindful towards their kneecaps.) The seat was equipped with a stabilising chute to prevent tumbling. Swedish Air Force doctors had studied the effects of body spin after ejection and contributed to the design.

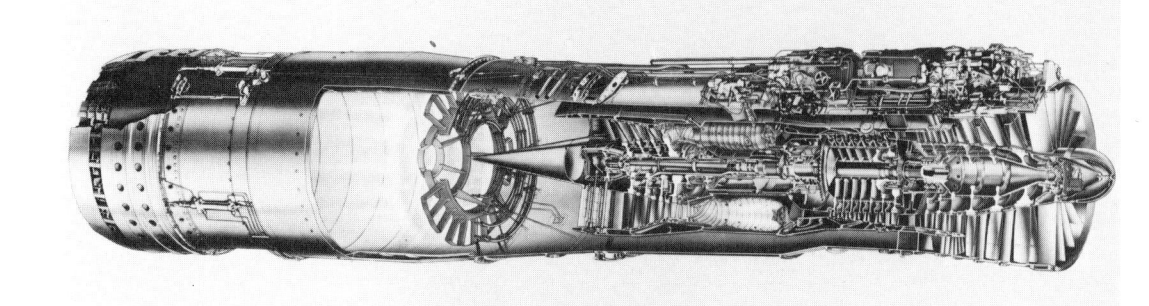

The Viggen's rocket-powered ejection seat was subjected to rigorous testing nearly three years before the Viggen itself flew. In May 1965, ground tests were begun. To study ejection characteristics at low speed, ejection was carried out from a truck using a real *Flygplan* 37 seat and a realistic dummy (wearing full flying gear). This gave SAAB engineers a fairly close approximation of conditions to be faced during takeoff and landing at speeds between 0 and 75mph (120km/hr).

It was decided early that the ground experiments, whatever their merit in testing 'slow speed' ejection problems, would not provide a realistic simulation of ejection problems to be encountered in high speed regimes. The Viggen ejection seat would have to be subjected to actual flight tests if it were to be proven out.

For this purpose, the SAAB company-operated second prototype J32B Lansen (FV serial number 32502) was fitted with a complete Type 37 ejector seat in the rear position behind the pilot. Actual ejections were made from the Lansen, again using well-dressed dummies but not real people at speeds up to 620mph (1,000km/hr).

Left:
A Swedish military version of the Pratt & Whitney JT8D turbofan engine, the Volvo Flygmotor RM8 combines low fuel consumption with high thrust. *Volvo Flygmotor*

Below left:
Technicians at the Volvo Flygmotor facility at Trollhattan make a test run of the RM8B turbofan engine used in the Viggen. *Volvo Flygmotor*

Below:
Volvo Flygmotor built a new test centre for evaluation of the Viggen's RM8 jet engine. The first test example of the RM8 is shown undergoing static run-up tests.
Volvo Flygmotor

Propulsion
After considering other possibilities as already noted, the Viggen's designers chose a version of the Pratt & Whitney JT8D-1, already thoroughly proven on the Boeing 727, the Douglas DC-9, the Super Caravelle and other types. The engine acquired the *Flygvapen* designation RM8, short for *reaktionmotor 8*, the RM8A being used on the AJ37 attack Viggen and its training and surveillance variants, the RM8B being used on the JA37 interceptor.

The RM8 was developed from the Pratt & Whitney design by Svenska Flygmotor (SFA) for production at its factory in Trollhatten. Since 1941 the Volvo car-manufacturing firm had been a majority shareholder in SFA and in the late 1960s the firm became Volvo Flygmotor. The programme was undertaken with co-operation from Pratt & Whitney which, in late 1963, supplied SFA with three special models of the JT8D-1 able to be fitted with an afterburner for testing. The engine consists of a bypass type gas generator and an afterburner. It provides good fuel economy for long range flights without afterburner

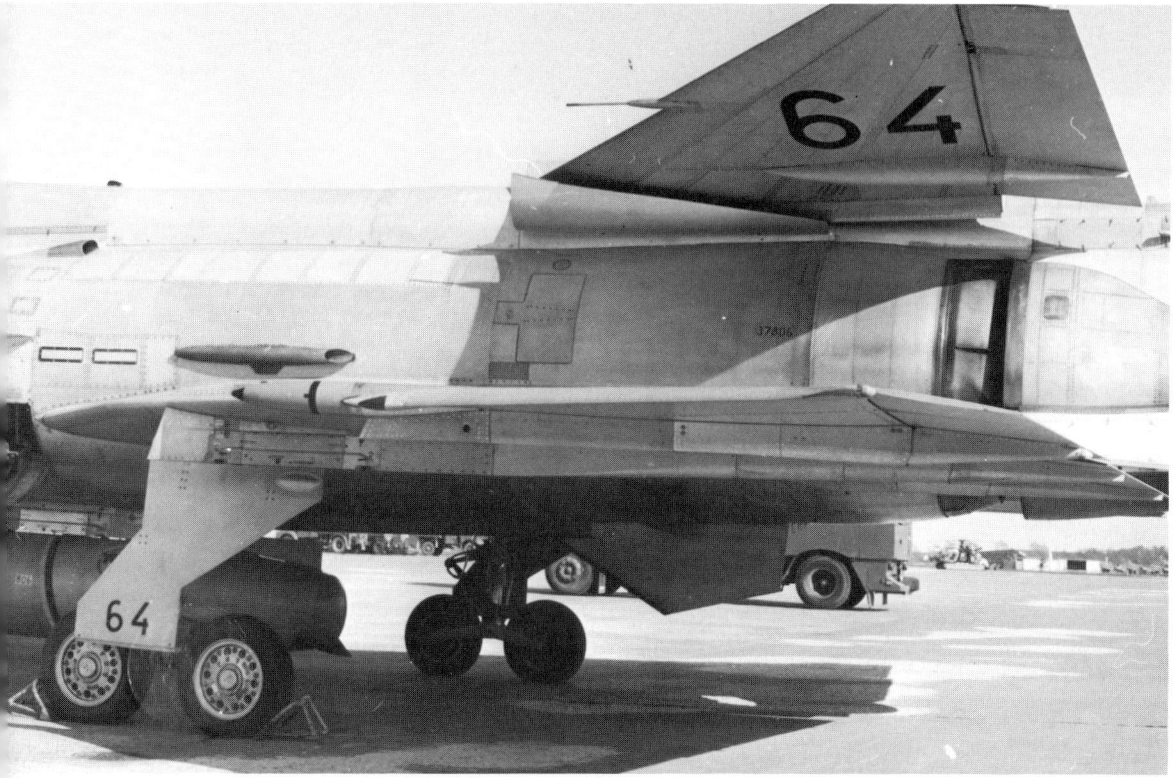
64
64

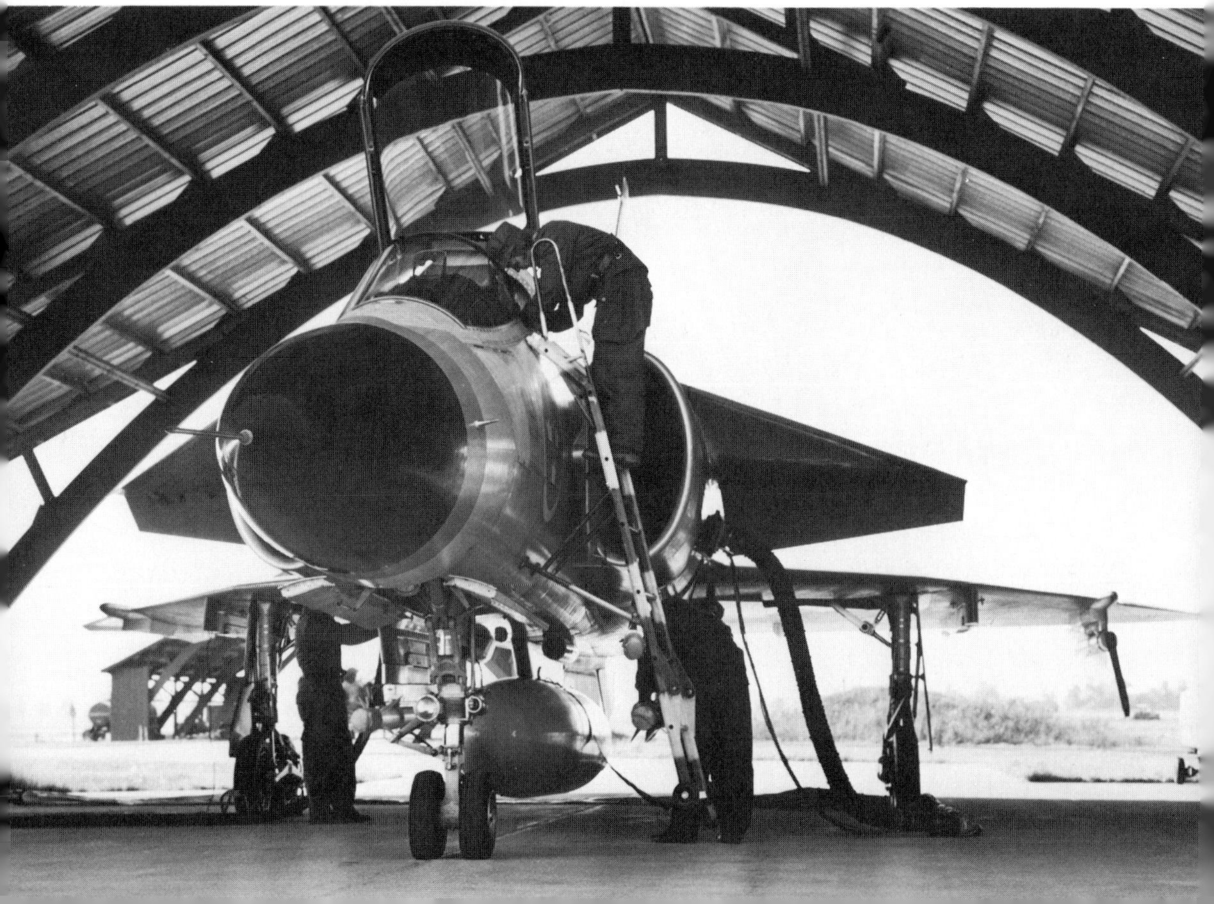

Above left:
For ease of handling and movement in and out of small hangars, the Viggen was designed with a folding vertical fin. *Saab-Scania*

Left:
In an 'alert' revetment, an AJ37 Viggen receives attention from mechanics. Although complex and sophisticated, Viggen was designed for ease of maintenance. *Saab-Scania*

Above:
Jet fighters customarily eat up vast amounts of fuel operating at low altitude. Still, Viggen was designed to fly and fight at low level as well as high. *Saab-Scania*

and very high thrust for take-off, climb and combat acceleration with afterburner. The thrust with afterburner may be changed continuously within three thrust levels.

The afterburner is mounted directly on the engine. Together they form a single unit which is easy to remove and reinstal. As a result of the bypass principle the outer engine structure remains relatively cool, reducing cooling needs and the risks of fire.

The air intake has been developed for good performance within a broad speed-altitude range and with simplicity of design.

Short-field requirements made it necessary to use a thrust reverser, which is located in the ejector at the aft end of the fuselage. The reverser lids are automatically actuated by landing gear compression and the engine jet is deflected forward through the annular ejector intake slot. The ejector intake is normally open at subsonic speeds in order to reduce fuselage base drag. At supersonic speeds, with the intake closed, the ejector serves as a supersonic nozzle.

Checkout and Monitoring of Electronic Equipment

The Viggen's developers realised from the outset that modern military aircraft must rely upon very costly electronic gear. SAAB designers wanted an airplane which would be safe, with reliable flight control systems and radio gear, and they also wanted an airplane which would be 'up' – available for service rather than down for maintenance. To assure reliability of the Viggen's electronic systems, the SAAB team created three firm objectives:

1 Flight safety would not be endangered by any single malfunction;
2 It would be possible to check the aircraft out on the ground before take-off so that 'alert' aircraft would be able to complete missions as planned and aircraft with faults could be quickly replaced;
3 It would be possible to service and maintain Viggen electronics rapidly and with limited personnel. This would be achieved through extensive automation of checkout and fault-tracing procedures.

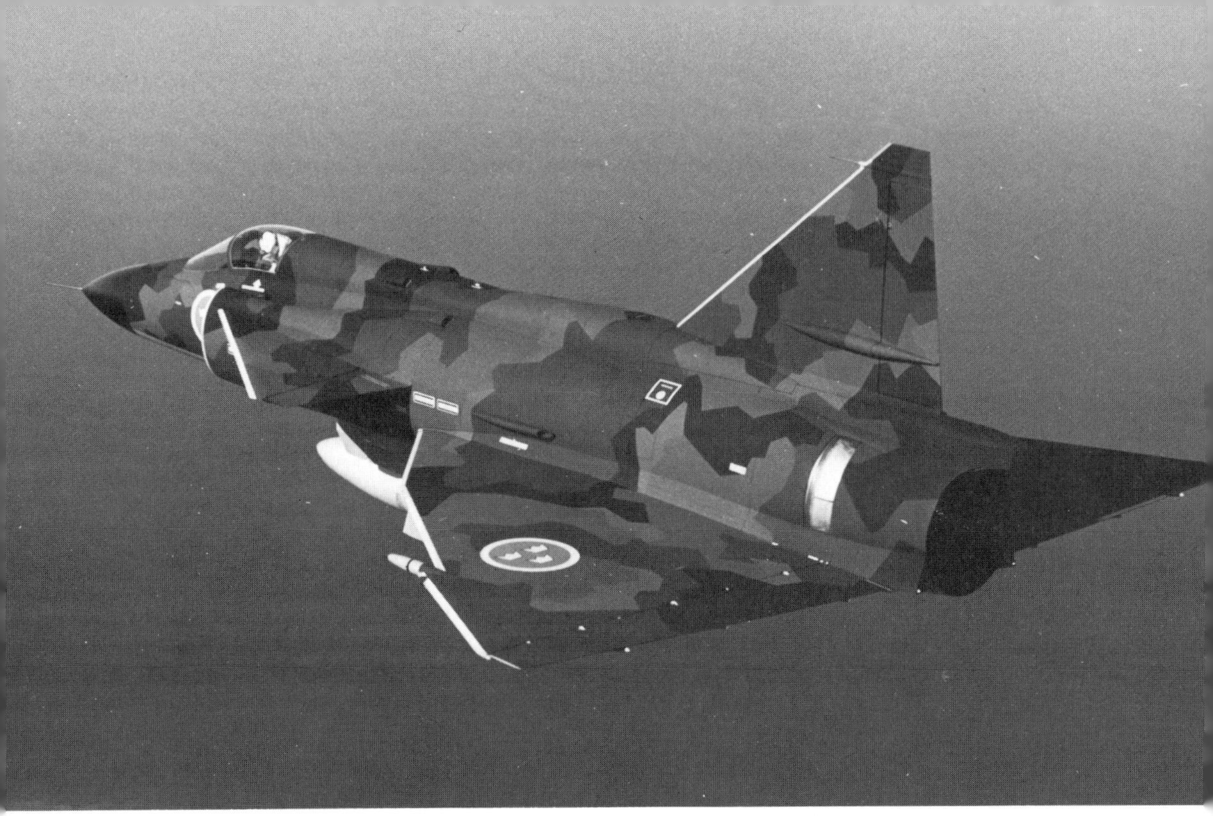

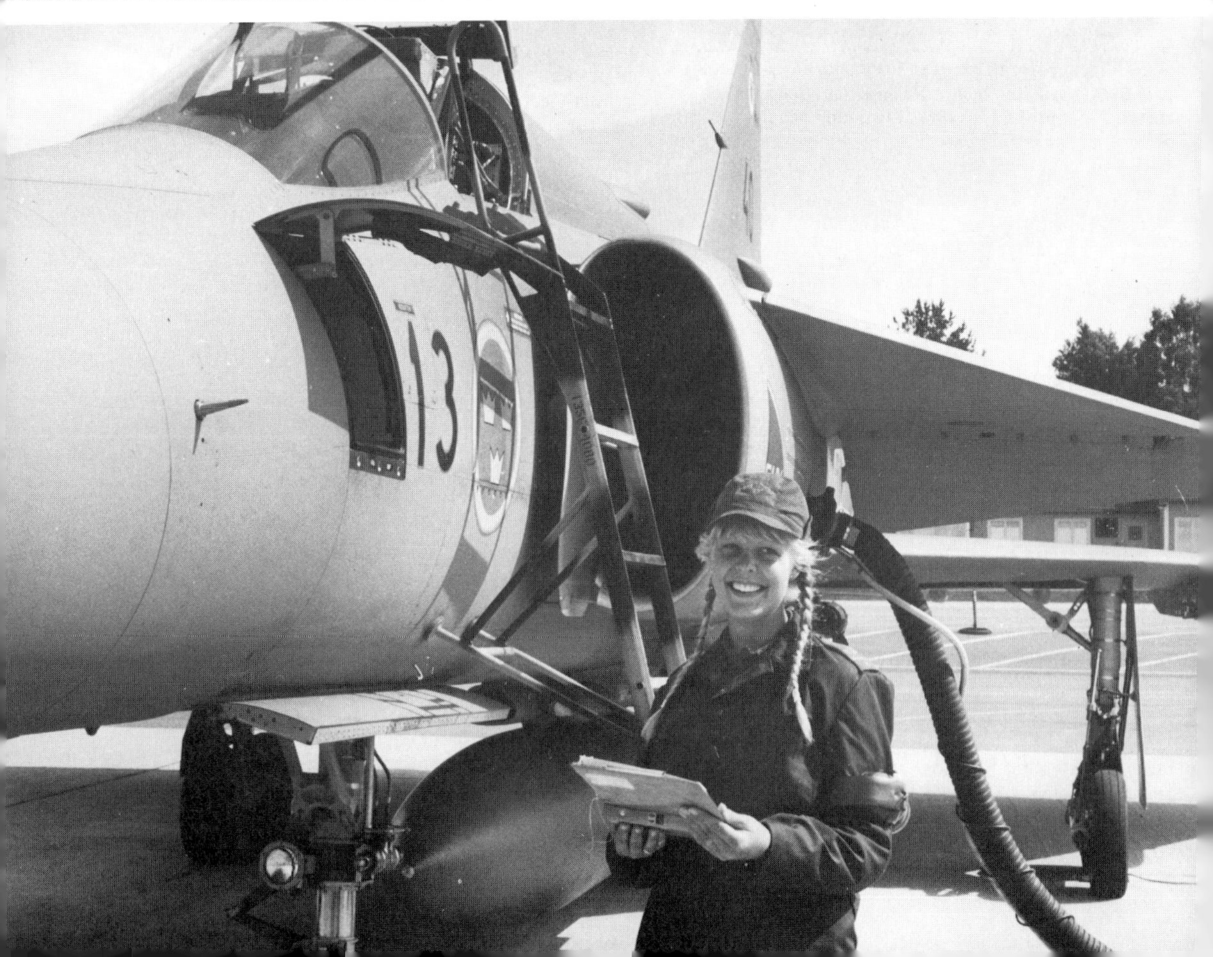
13

Above left:
By 1974 the Swedish Air Force was beginning to paint AJ37 Viggen aircraft in green/brown forest camouflage rather than leaving them in an easily-spotted natural metal. A decade later in 1984, the notion of camouflaging aircraft on the ground was to be abandoned and Sweden would adopt a drab grey paint scheme for air-to-air concealment. *Saab-Scania*

Left:
Not all Viggen mechanics are men. This JA37 interceptor, with easily accessible maintenance panels shown, is painted in the greyish air-to-air camouflage scheme adopted in 1984. *Swedish Air Force*

Above:
The canard foreplanes of the AJ37 Viggen are emphasised in this close-up of a row of aircraft at F7 wing, Satenas. The small horizontal strip beneath cockpit and foreplane is a formation light, and the word 'Fara' inscribed at air intake means 'danger' and warns 'groundlings' away from intake during engine run-ups.

SAAB designed a plane that keeps monitoring itself. In flight, all functions directly affecting flight safety are monitored continuously. Any malfunction is instantly flashed to the pilot in the form of a visual warning on a display panel. At the same time the faulty unit is put out of action. The pilot reverts to the standby system approved for every function that is vital to flight safety.

This self-policing aspect of the Viggen's design begins on the alert strip. While on cockpit alert, the pilot can initiate an essentially automatic checkout of all the electronic equipment in his aircraft, sub-system by sub-system. He reads the results on a checkout panel and – for certain checks – as instrument deflections. All the equipment necessary for this check is built into the aircraft, but if the checkout is performed without the engine running (which is usual) a small power unit is attached to the aircraft. Viggens at 'maximum alert', ie those ready to take off at a moment's notice, undergo checkout several times a day. The full checkout routine takes only a few minutes and can be interrupted instantly for immediate take-off if necessary.

If a fault is found in an aircraft, it is moved from its station near the runway to a special maintenance area which has test equipment containing all the instruments required for a complete performance test of the aircraft electronics. The equipment is mounted in two vehicles and can be connected to the aircraft at as many as 800 test points to perform up to 2,000 separate tests. This equipment performs most of the tests at a rate of one every one or two seconds, which is many times more fast than would be possible with non-automatc test equipment.

Most of the Viggen electronic equipment is connected to the airborne digital computer, and a special system has been developed for communications between the computer in the aircraft and the computer built into the test equipment.

6 Viggen Variants

Two generations of Viggen have spread their wings and vaulted into the skies. The first, retaining all of the key features of the original design, includes four variants: the AJ37 for ground attack, the Sk.37 two-seat trainer, the SH37 for maritime surveillance and the SF37 photo-reconnaissance machine. In a kind of rebirth, a new beginning which benefits from experience, the second-generation JA37 interceptor incorporates new radar, new avionics, an updated engine – and a gun. Also belonging to the second generation are the SAAB 37X export machine, the SAAB 37E Eurofighter and the A20 attack aircraft, none of which was ever built.

The following are the variations of Viggen.

The First Generation

AJ37
The AJ37 is, simply enough, a single-seat all-weather ground attack aircraft with a secondary fighter capability. The first AJ37, outwardly almost identical to the 37-1 to 37-8 prototypes, took to the air on 23 February 1971 for a 1hr 30min flight piloted by Ingemar Rasmusen, SAAB's chief of production flight testing. The first production example (37001) was rolled out in November 1970 and formally delivered on 21 June 1971, two weeks ahead of schedule, at ceremonies attended by Defence Minister Sven Andersson. The second machine (37002) was an

Below:
The unusual but far from unattractive lines of the AJ37 ground attack aircraft are evident in this study of Viggen 37064 at F6 wing, Karlsborg, touching down at RAF Coltishall on a rare visit to Great Britain on 18 September 1983. *John Dunnell*

Right:
A sequence of three photos shows camouflaged AJ37 Viggen attack aircraft (37076) turning on typical Swedish snow which renders the camouflage useless. Because tactical tail numbers end at 70, 37076 cannot wear 76 on its tail. It is numbered as aircraft 21 with F7 wing at Satenas. Photos taken February 1979. *Swedish Air Force*

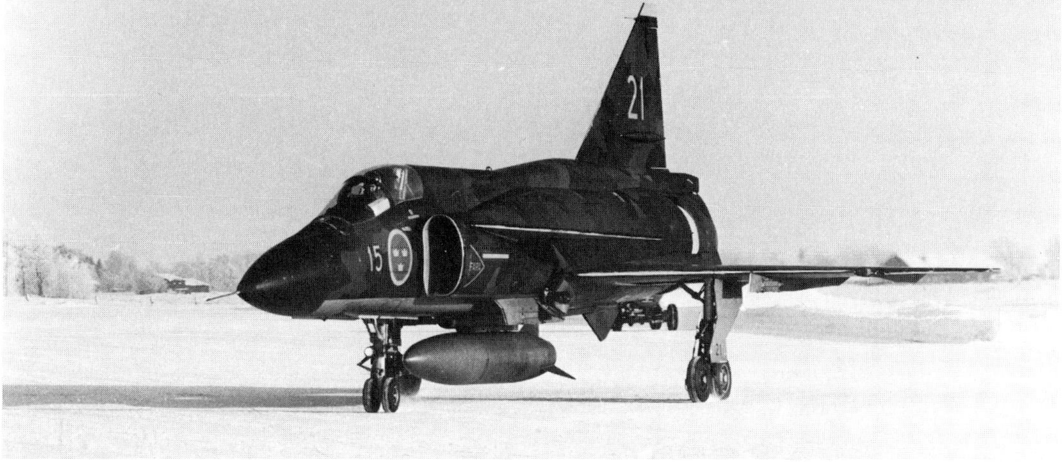
21
15

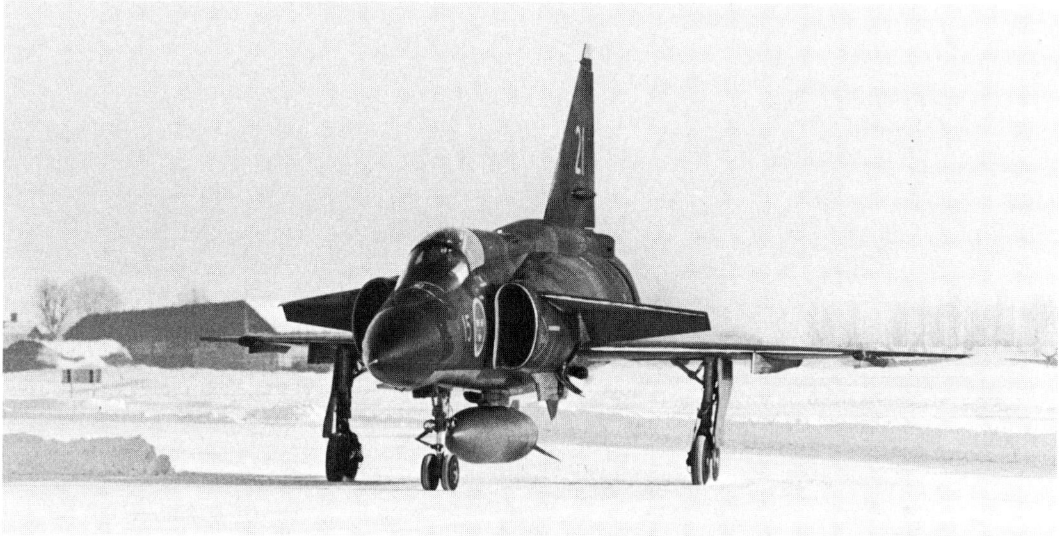
21
15

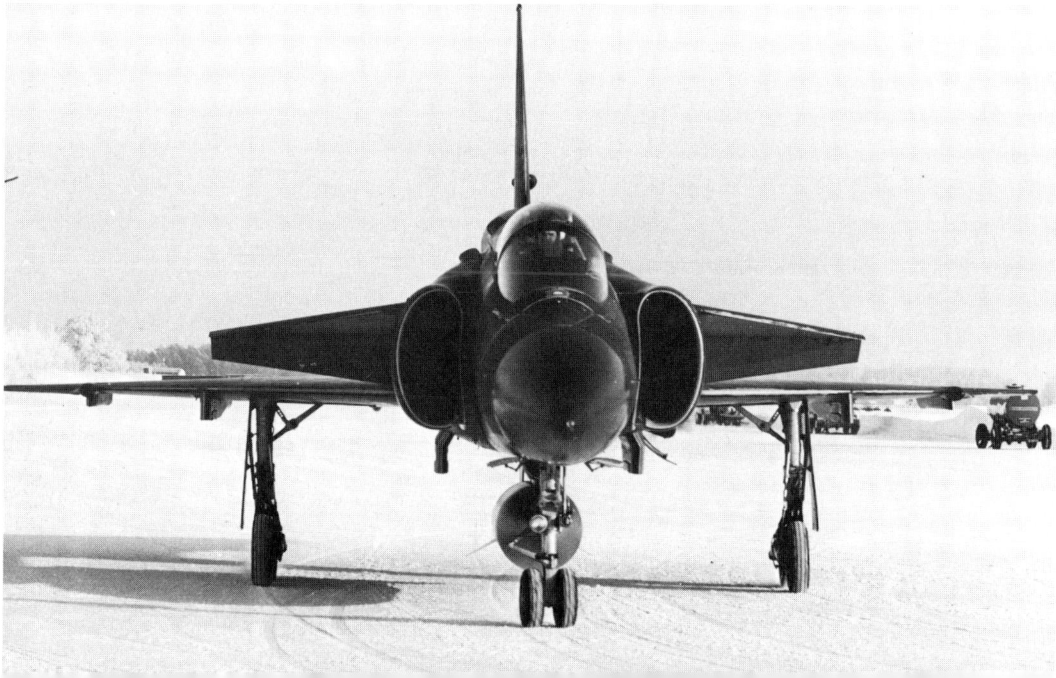

SWEDISH VIGGEN PROGRAMME PLAN

AJ 37 ATTACK VERSION

SK 37 DUAL COMMAND

SF 37 RECONNAISSANCE VERSION

SH 37 RECONNAISSANCE VERSION

JA 37 FIGHTER VERSION

AJ, SK, SF, SH ⟶

JA ⟶

74 75 76 77 78 79 80 81 82 83 84

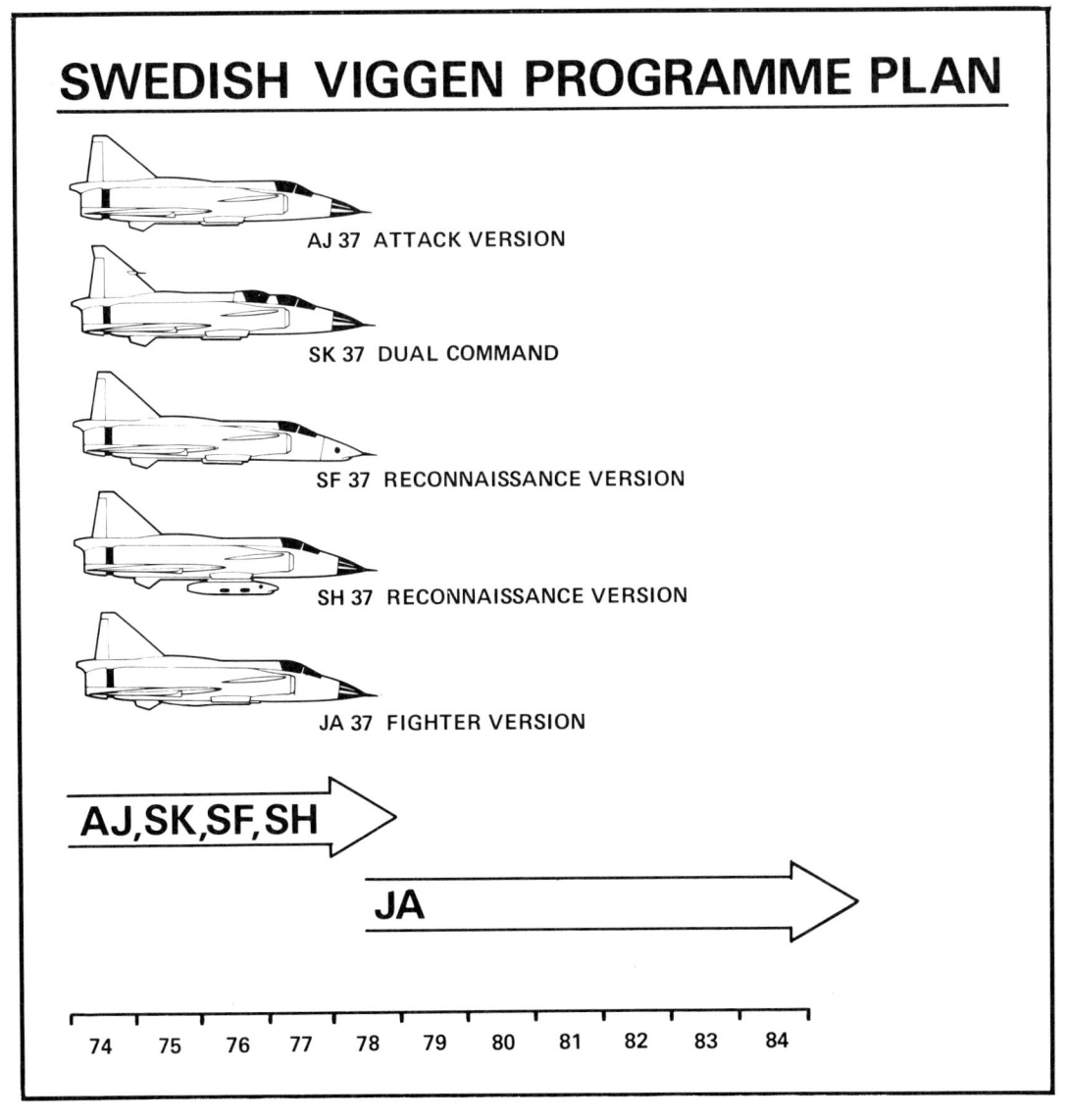

eye-stopper at the Paris Air Show that month. Delivered first to F7 wing at Satenas to replace the A32A Lansen, the AJ37 was outwardly virtually the same as prototype and development machines 37-1 to 37-8.

The aircraft is equipped with the advanced Ericsson radar described in the previous chapter, digital fire-control system with a Phillips air data computer, and a Marconi head-up display (HUD). Comprehensive navigational equipment includes Decca Type 72 Doppler radar, radar altimeter, a Tactical Instrument Landing System and a microwave scanning-beam blind-landing system. Extensive radar warning and ECM equipment is carried. The aircraft has seven hardpoints for weapons pylons which give a load capability of 15,432lb (7,000kg) of stores.

Above:
The five principal Viggen variants are shown together with a time frame for their entry into operational service. It should be understood that the JA37 fighter-interceptor variant is really a new generation of Viggen – in most respects a new airplane – advanced beyond the first four variants. This rough drawing does not show the distinctive difference in the upper tail shape of the JA37. *Saab-Scania*

Weapons normally used with this version include the SAAB Rb.04E anti-ship homing missile, the SAAB Rb.05A supersonic air-to-surface missile, the Hughes AGM-65 television-guided Maverick, and Rb.24 (AIM-9) Sidewinder or Rb.28 (AIM-4) Falcon air-to-air missiles. A 30mm Aden gun pod is available, as are a variety of bombs and pod-launched rockets.

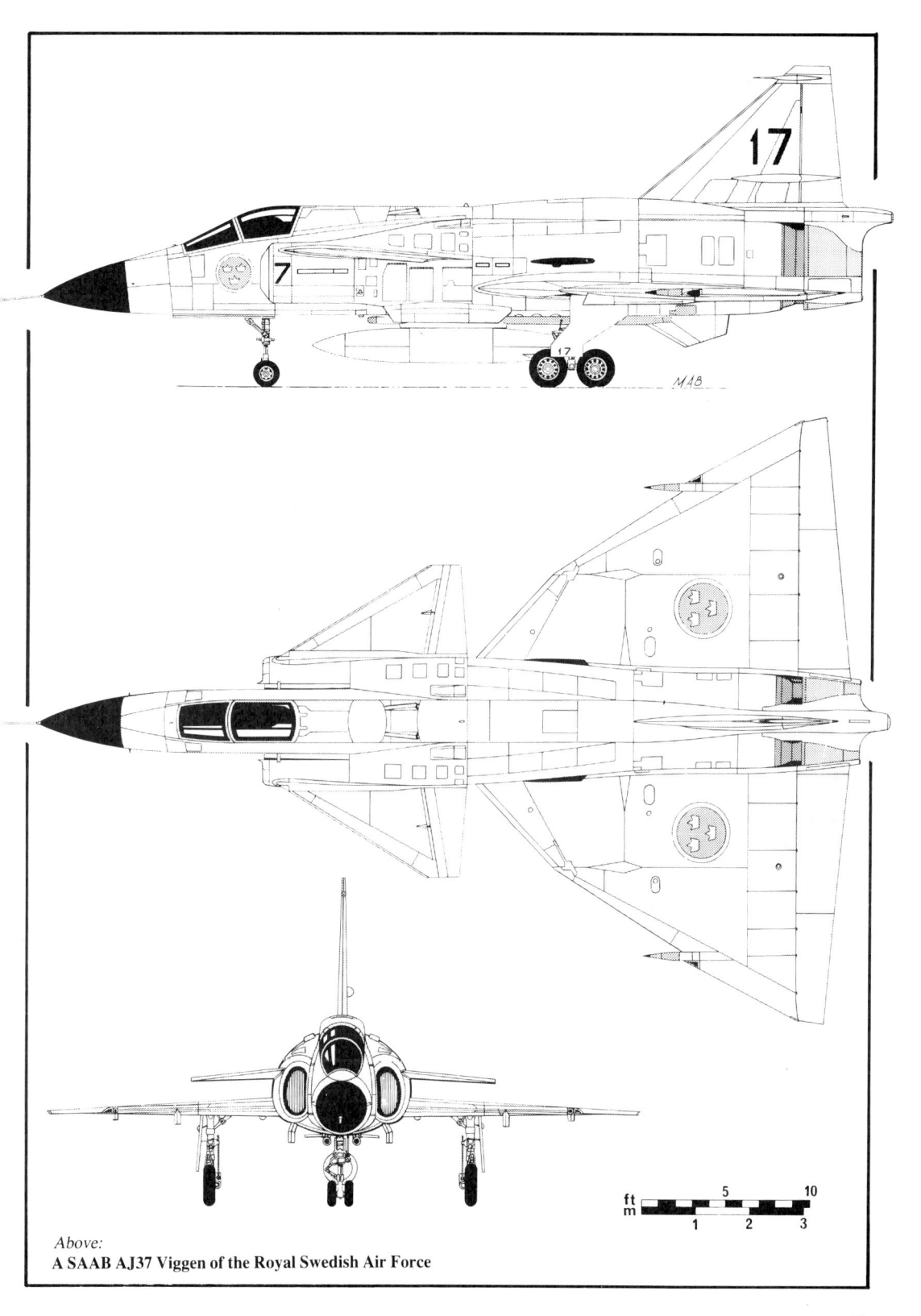

Above:
A SAAB AJ37 Viggen of the Royal Swedish Air Force

Units operating the AJ37 include F6 at Karlsborg (two squadrons), F7 at Satenas (two squadrons) and F15 at Soderhamn (one squadron).

It is understood that 175 examples of the AJ37 model were followed by an additional five machines, apparently covering the serial number range 37001 to 37180. The last time figures on Viggen losses were made available was in 1979 when it was reported that 21 aircraft, among them 20 of the AJ37 model, had been attrited, no fewer than six during the period May 1978 to July 1979. Known serials of aircraft lost, with attrition dates, include 37011 (10 October 1975); 37014 (6 October 1975); 37018; 37039 (11 January 1979); 37071 (27 August 1975), 37095 (also on 11 January 1979 in a mid-air collision). The dilemma faced by the aviation buff/historian is illustrated by the numerous sources which report that 37018 was lost in an accident on 31 April 1974. This item of information is firmly implanted in many spotters' records, even though there was no April 31 in 1974 or any other year!

Though the earliest AJ37 has now been in service for some 15 years, these Viggens have relatively low airframe hours for such long service. They are likely to have at least another decade's active life ahead before the JAS39 Gripen becomes predominant on the Swedish scene in the 1990s.

Sk.37

The Sk.37 is the two-seat dual control trainer Viggen with a secondary capability as an attack aircraft. Apart from the second seat, this model can be distinguished by a taller vertical fin than other first-generation Viggens. The first example of the Sk.37 (37800) flew on 2 July 1970 on a 70min flight with Per Pellebergs at

Below:
Two-seat Sk.37 Viggen tandem trainer No 37802. coded 68 of F7 wing at Satenas, takes off carrying the centreline fuel tank which is a permanent fixture of the twin-seat model. The rear-seat instructor has periscopes to aid with forward vision. *Saab-Scania*

Bottom:
Tow bar attached to nosewheel, a Sk.37 trainer shows the folding vertical tail which makes Viggen fit more easily into confined spaces. *Saab-Scania*

Left:
Sk.37 Viggen trainer No 37816 taxies out at a base belonging to F15 wing at Soderhamn in 1976, after camouflage was applied to most two-seaters. *Saab-Scania*

Below:
Making a rare visit to RAF Coltishall on 18 September 1983, two-seat Sk.37 Viggen (37813) touches down. *John Dunnell*

the controls. The Sk.37 entered service in June 1972 with the Viggen Conversion Unit of F15 wing at Soderhamn.

The Sk.37 provides for pilot conversion and supersonic training on Viggen in the *Flygvapen*. This version is also equipped for combat flying, if need be. The extra cockpit reduces fuel load and some avionics, so Sk.37s always fly with a large fuel tank mounted permanently on the fuselage centreline pylon. Twin forward-looking periscopes are provided for the instructor in the rear seat, and the extra side area of the aircraft is countered by the already mentioned fin-tip extension and a slightly enlarged ventral fin.

The foreign press was given its first close look at Viggen capabilities in May 1974 when Robert F. Ropelewski of *Aviation Week and Space Technology* and Hugh Field of *Flight International* were provided with orientation flights in the two-seater. Capt Bo Hellstrom, who flew both editors, said afterward that they were 'suitably impressed'. The flights in aircraft 37802 of F7 wing at Satenas represented a major effort to make the Viggen better and more favourably known outside Sweden's borders and came at a time when there still seemed hope the type might be selected for NATO air forces (see chapter eight).

Sk.37 two-seaters operate with F7 wing at Satenas and F15 wing at Soderhamn. It is understood that 18 Sk.37 airframes were completed, with serials 37800 to 37817. One Sk.37, serial 37806, serving with F15, was lost in an accident on 26 June 1980.

SH37

The SH37 is a single-seat, all-weather maritime reconnaissance aircraft with a secondary ground attack role, and has replaced the S32C Lansen. The first SH37 (37900) flew in June 1975 and appeared at the Paris Air Show that month. The SH37 is equipped with a nose-mounted surveillance radar with camera equipment to record the display. A long-distance camera is carried in a pod on the right-hand fuselage pylon and the usual range of reconnaissance pods, ECM pods, fuel tanks and missiles is carried on the remaining pylons. The SH37 is in service with one squadron of F13 wing at Norrkoping and together with the SF37 in mixed squadrons of F17 wing at Ronneby and F21 at Lulea.

Twenty-six SH37 airframes were manufactured, with serials 37900 to 37926. One is known to have been lost, serial 37906, which crashed near Soderhamn on 26 June 1980.

Below:
Photographed in mid-air by Swedish photographer Robert Lofberg in 1976, this SH37 sea-surveillance Viggen (37908) of F13 wing at Norrkoping prowls the Scandinavian coastline.

Top:
The SH37. Carrying out its sea-surveillance role, the SH37 Viggen banks in flight over a naval vessel in the congested Baltic Sea. *Saab-Scania*

Above:
The SH37. Depicted in 1976 shortly after deliveries began; a SH37 Viggen of F13 wing at Norrkoping flies at low level over a Swedish lake. *Saab-Scania*

Below left:
The SH37. A formation of Viggens, with a SH37 maritime surveillance aircraft (37908) in the foreground. *Saab-Scania*

SF37

Development of a photo-reconnaissance Viggen was assured in May 1971 when the Swedish government authorised the Defence Materiel Administration to proceed with an initial expenditure of Sw Kr61million (£5million). Further funding preceded roll-out of the first airframe. The SF37, the single-seat Viggen for all-weather day and night photo-recconnaissance, first flew in May 1973 and has replaced the S35E Draken.

Lacking the attack radar of the AJ37, the SF37 has a different nose section housing low and high level optical cameras and infra-red equipment and sighting systems. The outer fuselage pylons are normally fitted with reconnaissance pods, of which two are in wide use: the Red Baron pod housing an Infra-Red Linescan System (IRLS) and optical cameras, and a night illumination and camera pod. The central pylon is usually used for a fuel tank, while the wing pylons are capable of carrying both ECM pods and Rb.24 missiles for self-defence. The SF37 is in service with F13 wing at Norrkoping, F17 at Ronneby and F21 at Lulea.

It is thought that 28 SF37s were built, bearing serials 37950 to 37977. The final example of a first-generation Viggen, the last SF37, was completed at Linkoping on 1 February 1980 and delivered to F21 wing at Lulea – clearing SAAB's production lines for a virtually new machine with totally new potential.

75

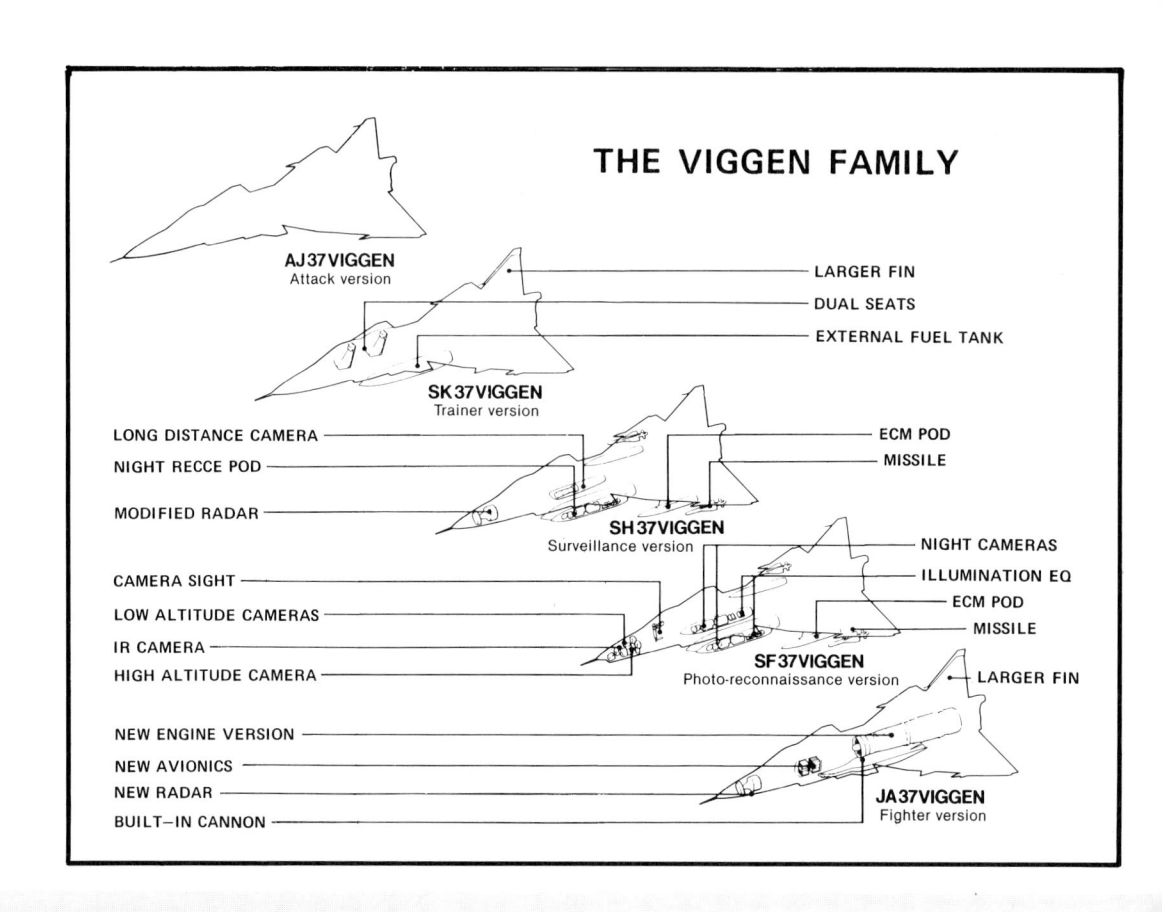

THE VIGGEN FAMILY

AJ 37 VIGGEN
Attack version

SK 37 VIGGEN
Trainer version
- LARGER FIN
- DUAL SEATS
- EXTERNAL FUEL TANK

SH 37 VIGGEN
Surveillance version
- LONG DISTANCE CAMERA
- NIGHT RECCE POD
- MODIFIED RADAR
- ECM POD
- MISSILE

SF 37 VIGGEN
Photo-reconnaissance version
- CAMERA SIGHT
- LOW ALTITUDE CAMERAS
- IR CAMERA
- HIGH ALTITUDE CAMERA
- NIGHT CAMERAS
- ILLUMINATION EQ
- ECM POD
- MISSILE

JA 37 VIGGEN
Fighter version
- NEW ENGINE VERSION
- NEW AVIONICS
- NEW RADAR
- BUILT-IN CANNON
- LARGER FIN

Above left:
A diagram depicting major differences among Viggen variants. *Saab-Scania*

Left:
The SF37 reconnaissance Viggen. Aircraft number 02 of F13 wing on the ground at Norrkoping with auxiliary power unit (APU) connected, ready for a mission. *Swedish Air Force: Rune Rydh*

Above:
The SF37. A camouflaged reconnaissance Viggen with a full load of fuel and sensors lifts off from Norrkoping for a tactical mission. *Swedish Air Force: Rune Rydh*

Right:
The SF37. The ease of access to the optical camera systems of the SF37 reconnaissance Viggen is illustrated by this close-up view of ground crew members loading camera gear in the forward nose bay. On return from a combat mission cameras can be unloaded and film processed and prepared for analysis in a matter of minutes. *Swedish Air Force: Rune Hedgren*

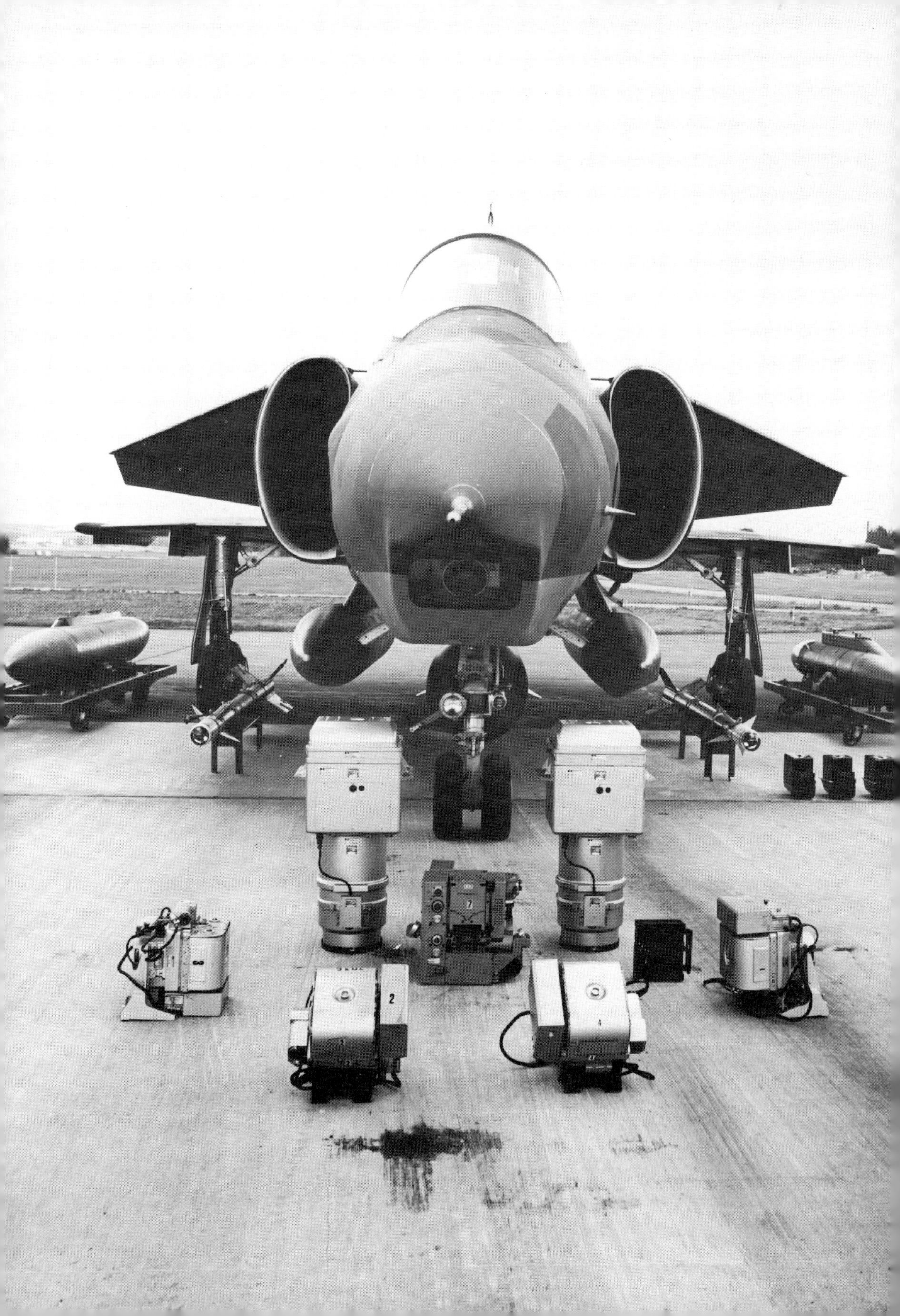

Left:
Viewed from straight on, the SF37 reconnaissance variant of the Viggen is depicted with the diverse camera systems it can carry. *Swedish Air Force: Rune Rydh*

Below:
The SF37 reconnaissance aircraft, connected to an auxiliary power unit, is ready for a mission. *Swedish Air Force: Rune Rydh*

The Second Generation

JA37

Though the future of a second-generation Viggen interceptor seemed clouded at times, in March 1970 the Swedish Air Defence Committee recommended the go-ahead in a report to Supreme Commander Gen Torsten Rapp. This committee, appointed in 1967 and headed by Maj Gen Bo Westin, in effect cleared the way for the JA37, the Jaktviggen.

The first aircraft configured as a JA37 interceptor was flown on 15 December 1979 after being preceded by four developmental machines (see below). The JA37 is a single-seat, all-weather interceptor with the improved Volvo Flygmotor RM8B engine, with greatly improved performance. It retains a secondary capability for the ground-attack mission.

Flight testing of selected systems, including the improved radar described in the previous chapter, began in early 1973 in a modified Lansen development aircraft. Four of the initial SAAB 37 developmental machines were used in the JA37 programme. The first of these was the second Viggen built, 37-2, renumbered as 37-21, retaining features of the earlier craft but with the revised control system of the second-generation interceptor. Viggen 37-21 was first flown in this configuration on 4 June 1974.

Second of the four developmental machines, used for RM8B engine tests, was the seventh Viggen built, 37-7, renumbered as 37-71 and flown on 27 September 1974. This aircraft carried the 30mm Oerlikon cannon installed in an underbelly pack and had the revised ventral fin characteristic of the JA37.

The third developmental machine, the third Viggen 37-3 now renumbered 37-31, flew on 22 November 1974. A fourth machine followed before the first aircraft built from the outset as a JA37 interceptor, using JA37 factory tooling, 37-8, flew on 15 December 1979. Like all production JA37s to follow, this craft had four elevon hydraulic actuators under each wing instead of three on earlier Viggens and had the modified, taller tail-fin already found on the Sk.37.

The JA37 may, to the untrained eye, appear almost identical to other Viggens. In fact the Jaktviggen, as the second-generation interceptor is often called, is optimised for the fighter mission with a refined and reinforced airframe. In addition to the Oerlikon cannon already described, the JA37 carries six British Aerospace Sky Flash semi-active radar homing air-to-air missiles designated Rb.71 by the *Flygvapen*. JA37 aircraft eventually will also carry the Swedish version of the AIM-9L Sidewinder, the first infra-red homing missile which can be used in a frontal attack as proven in combat with Sea Harriers in the Falklands.

A less important but still dramatic change in the Jaktviggen is an entirely new cockpit presentation system significantly improved over that of the AJ37, itself also highly praised, developed by Saab-Scania in concert with Svenska Radio. The system is built up around three main electronic displays providing flight information: the head-up display (HUD) developed by Smiths Industries in collaboration with Svenska Radio for low-level flight and close-in combat; the head-down display (HDD) developed by Smiths Industries for all-weather interception, and the tactical display for en-route navigation or combat situation assessment, embodying a synthetic video map. ('Being in the cockpit is a little like playing Star Wars', grins one pilot.) All three displays can be read by the pilot even in bright ambient light.

The first unit to employ the JA37 operationally was F13 wing at Norrkoping. Scheduled to receive the aircraft are F17, F1, F4, F10, F16 and F21. JA37s were initially delivered in the familiar, multi-hued, green/brown forest camouflage long familiar to Viggen watchers and intended to conceal the aircraft on the ground. Since advanced sensors now render ground camouflage almost useless, JA37s are now being painted in a low-visibility grey for concealment in an air-to-air combat setting.

Below:
The second-generation JA37 or Jaktviggen interceptor, Aircraft 37-8, seen here, was the first developmental machine built from the start with JA37 tooling. It is carrying dummy Sky Flash missiles.
Saab, via Michael France

Bottom:
The first fully developmental Jaktviggen or JA37, aircraft 37-8, is seen in natural metal on a 1975 test flight.
Saab, via Michael France

Above:
The JA37 interceptor. This Jaktviggen, aircraft 39 of F13 wing at Norrkoping (37302), climbs aloft with fuel and missile load. *Saab-Scania*

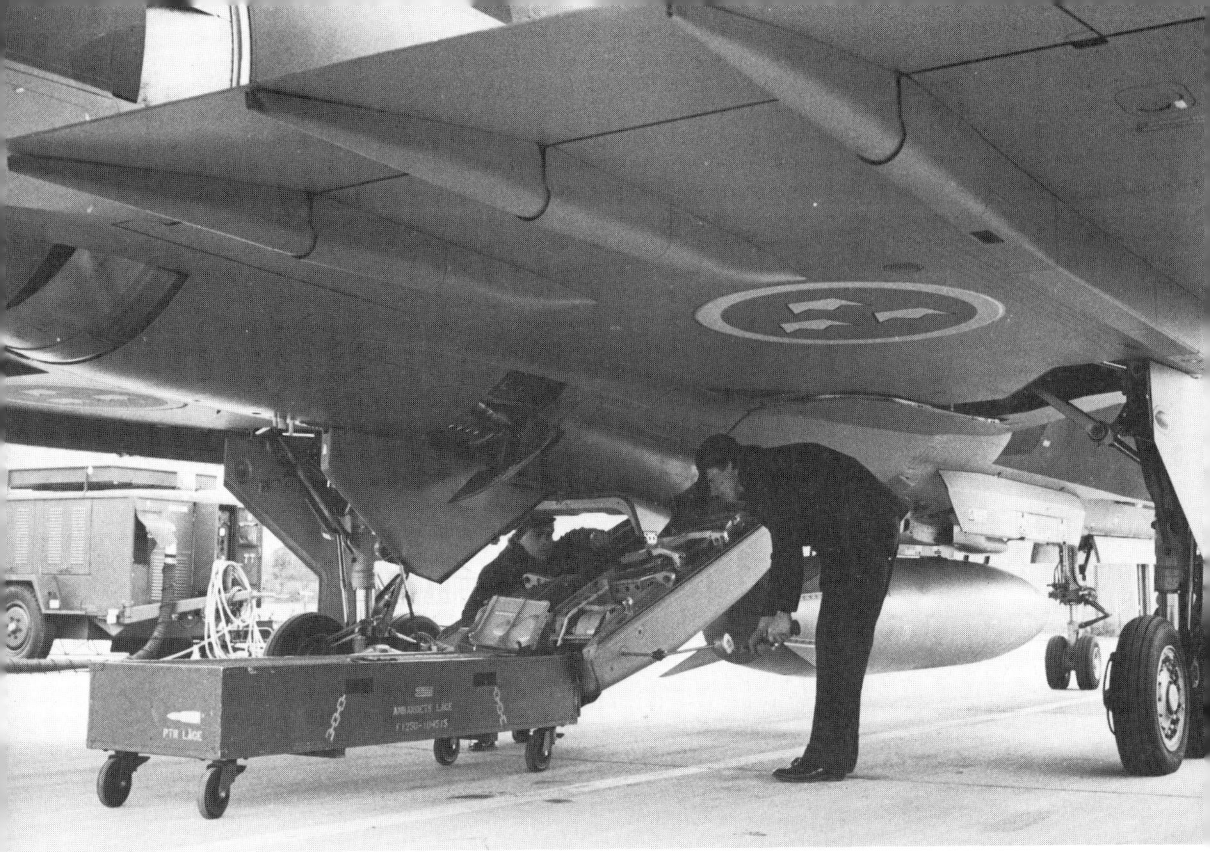
AMBANDETS LÄGE
F1250-104515

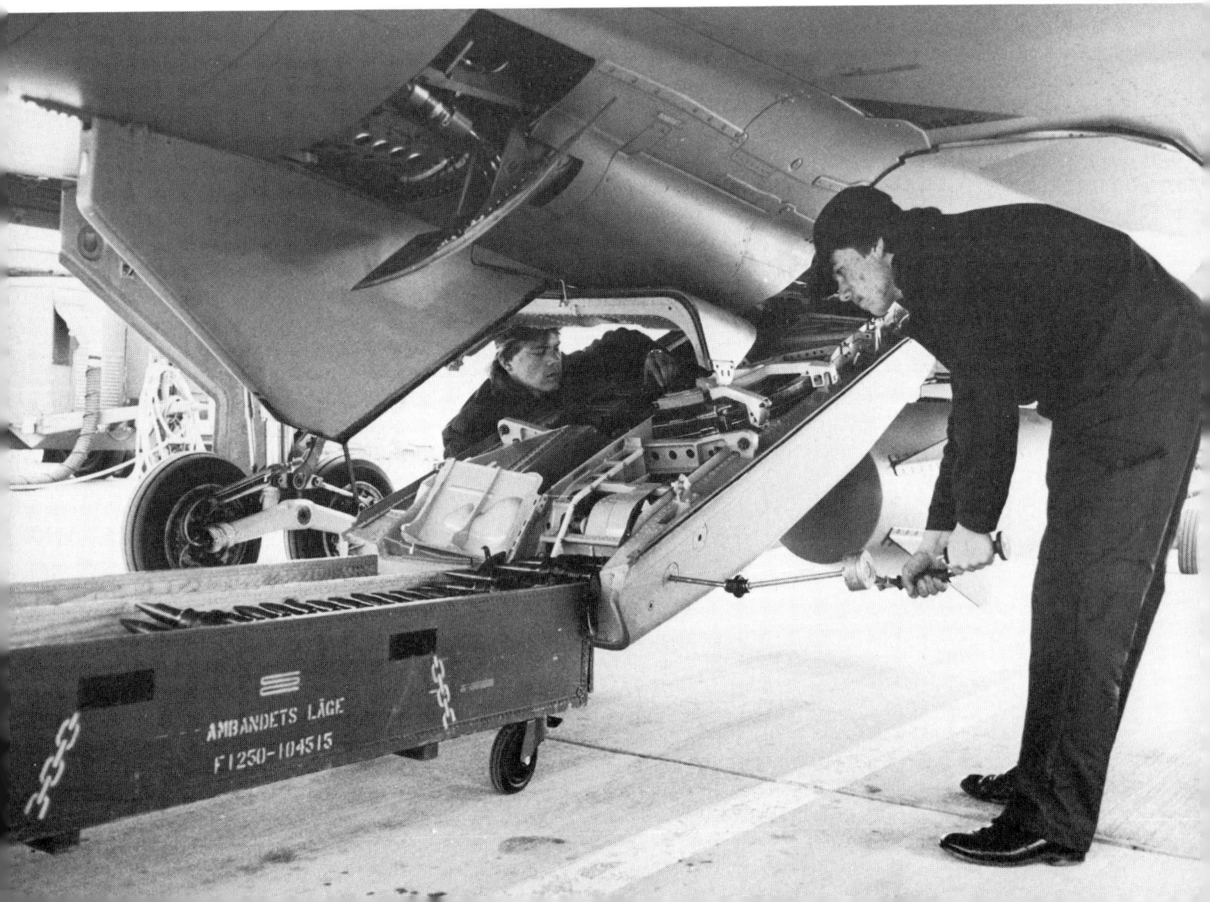
AMBANDETS LÄGE
F1250-104515

This series of close-ups of the operational JA37 interceptor Viggen shows ground crews loading its 30mm Oerlikon cannon (*Left*), installing Rb.24 Sidewinder missiles (*Above*), and installing the Rb.71 Sky Flash missile (*Right*). All were taken in 1984 at F13 wing at Norrkoping.
Swedish Air Force: Rune Rydh

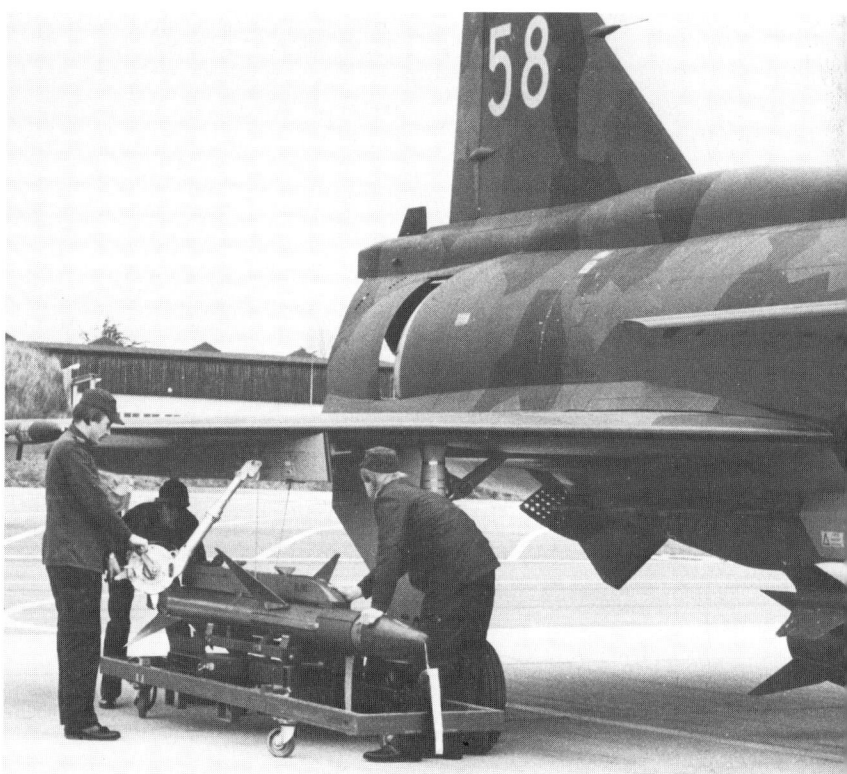

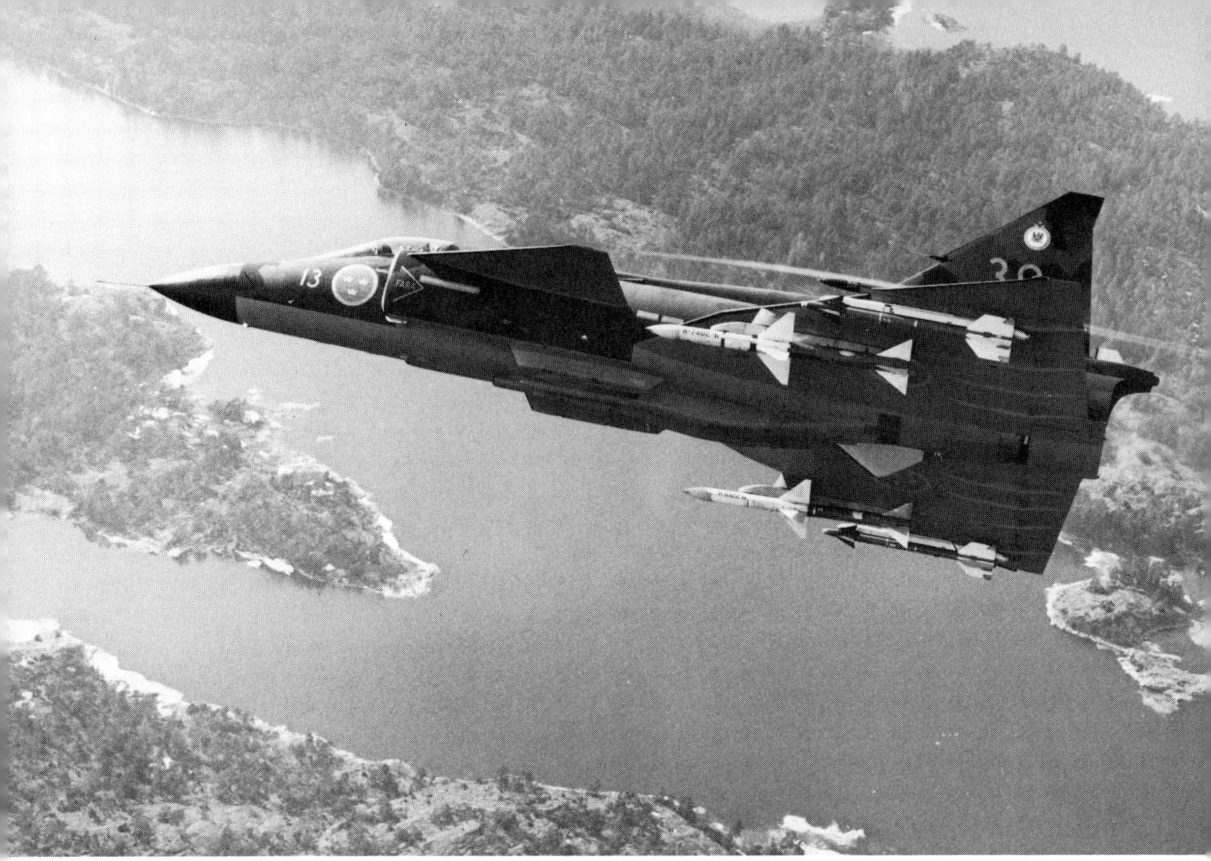
13
FARA

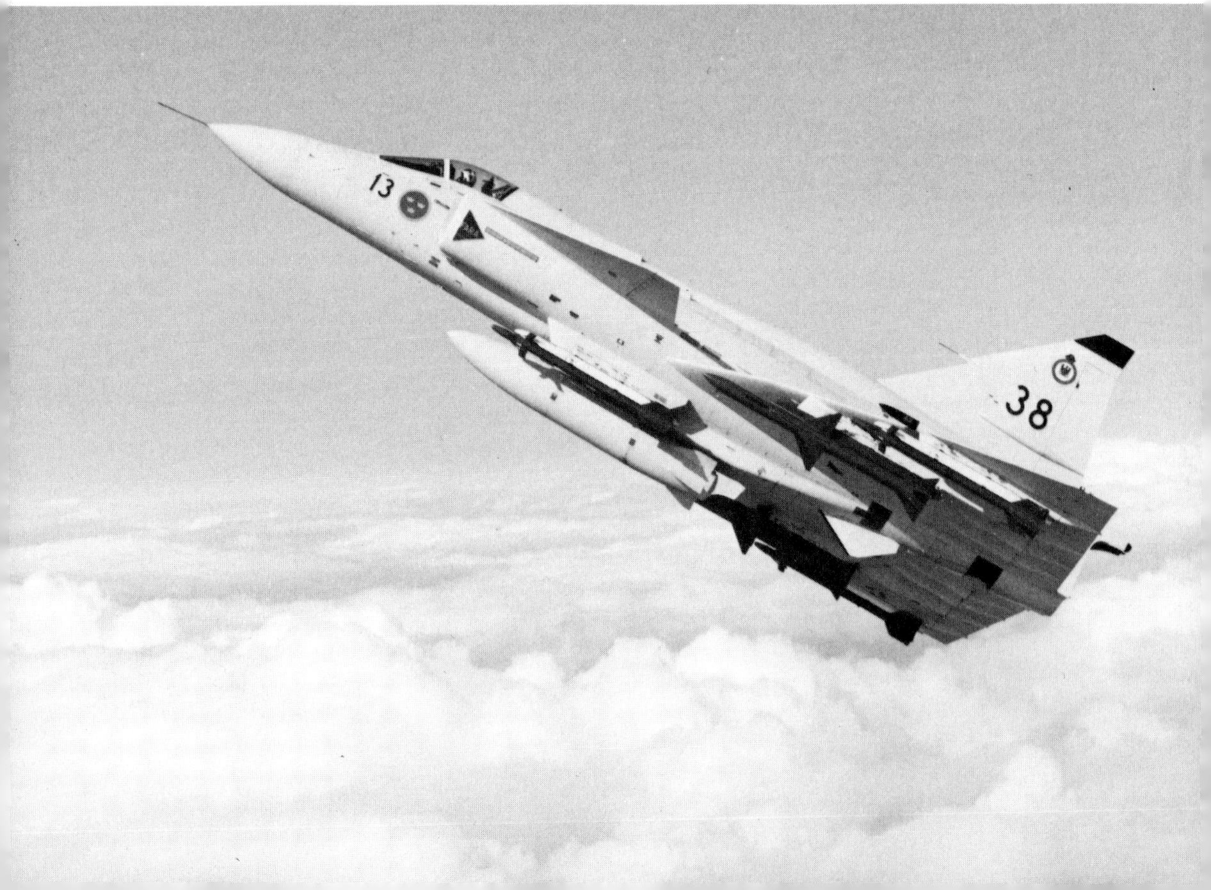
13
38

Above left:
JA37 interceptor 37302 peels over a Swedish lake carrying underwing British Aerospace Sky Flash missiles.
Saab-Scania

Left:
By 1982 the Swedish Air Force had determined that modern technology made it impractical to camouflage aircraft on the ground and sought, instead, a low-visibility camouflage for air-to-air combat. This JA37 interceptor Viggen was painted in a snow-white camouflage scheme which seemed promising, but was not adopted. It is thus a 'one of a kind' Viggen, seen here with centreline fuel tank and mixed armament of Sky Flash Sidewinder missiles.
Saab-Scania

Top:
The unique, all-white JA37 interceptor, aircraft 38 of F13 wing at Norrkoping, in a low-level go-around.
Swedish Air Force: Runc Rydh

Above:
A truly superb and exquisitely detailed scale model of the 'one of a kind' all-white JA37 interceptor is captured in a pose showing how this second-generation Viggen might look in readiness for a night mission. The model was made from an ESCI 1/48th scale kit, one of several Viggen kits available to modellers. *Wolfgang Leipe*

Above:
Viggen family portrait: for the first and possibly only time in one photograph, all five variants of the Viggen fly together. From left to right: AJ37 attack, Sk.37 trainer, SH37 maritime surveillance, SF37 photo reconnaissance and JA37 second-generation fighter interceptor. *Saab-Scania*

Below:
A JA37 Viggen interceptor in flight. *Saab-Scania*

SAAB 37X

The designation SAAB 37X was applied to a proposed export-fighter Viggen of the late 1960s, at a time when SAAB was pondering foreign sales of the aircraft but long before the 'deal of the century' which resulted in NATO nations acquiring the General Dynamics F-16. The SAAB 37X proposal is believed to have been shown to Great Britain and Norway (see chapter eight) among other nations, but never progressed beyond the early planning stage.

SAAB 37E

The SAAB 37E Eurofighter was the export version of the Viggen proposed for the NATO nations when they settled, eventually, on the F-16 instead. Discussions of the proposal were held at the Hanover air show in April-May 1974. In August 1974 SAAB formally proposed the Eurofighter as a NATO replacement for the Lockheed F-104. The Eurofighter would have been made available under a financially attractive scheme whereby subcontract work would be licensed overseas for *all* Viggens, including those operated by Swedish units, while the customer country would also have benefitted from the setting-up of a Volvo truck factory or a Saab-Scania avionics or subsidiary industry. The Eurofighter would have been a lightweight version of the JA37. Its design proceeded much farther than that of the SAAB 37X, but it never flew.

A20

A20 was the out-of-sequence designation assigned to a medium-range attack variant of the second-generation Viggen which would have been built simultaneously with the SAAB B3LA attack aircraft. The latter appears to have been a low-cost airframe intended to reconcile the wish for home-built aircraft with the rising prices of modern fighters. Both the A20 and B3LA were cancelled at an early stage when Sweden explored foreign designs, such as the McDonnell F/A-18A Hornet, before settling on the JAS39 Gripen as its future fighter.

7 Flying the Viggen

Above:
Sweden is world-renowned for its beautiful airplanes. Here, a JA37 Jaktviggen is checked and serviced by a female conscript performing duty as part of a training programme leading to an officer's commission. The JA37 in the background wears the low-visibility grey paint scheme which is becoming standard. *Swedish Air Force*

The Viggen pilot wakes up to begin his day. His is a comfortable setting, perhaps unique among the world's air forces. No tent, no Quonset, no transient Butler-barrack accommodates him. If he is a senior officer like Lt Col Erik Svensson, commander of the 1st *Attackflygdivision* (attack squadron) of F6 wing at Karlsborg, he may set forth from his family's three-bedroom bungalow on base where wife and children reside permanently with him in the high standard of living enjoyed by Swedish citizens in all walks of life. If he is one of the younger men the Air Staff wants at the controls of its fighters, like Lt Goran Andersson of F13 wing at Norrkoping, he will start toward the morning briefing from a sumptuously furnished Bachelor Officers Quarters which is, in essence, a permanent home. To the extent possible, *Flygvapnet* keeps men in the same wing, at the same base, to encourage stability and teamwork and to promote a sense of identity with the unit. In some other air force a fighter pilot may wake up grappling for his suitcase, gawking at bare unfamiliar walls, pondering local currency or language, finding that he forgot the razor blades, groping to remember the whereabouts of chow hall, briefing room, base operations. While many fighter pilots deploy vast distances, to any climate, under any conditions, at times with little more than flight suit and toilet bag, the Viggen flier's sole purpose is to defend his homeland so he benefits from a sense of continuity uncommon in the profession of aerial arms. He knows every nook and cranny of his home base, and usually returns to it from a mission. He has been to every other fixed airfield in Sweden often enough that he knows them well, too. He is on a first-name basis with an extraordinarily large portion of his fellow officers in *Flygvapnet*, including superiors in a society where informality is the norm. Yes, there are deployments to remote roadway bases. Yes, there are manoeuvres which take him to unfamiliar billets. But runway heading, tower locations, the right choice of cuisine in the mess, the phone number for base operations, all are cemented permanently in his mind.

He heads for the briefing on foot, or using his own car or bicycle. He may have occupied the same cushioned chair in the briefing room for months, even years. The faces around him – briefer, crewchief, wingman – are familiar. He cannot predict which machine he will fly on a given day, though: Viggens are not assigned to individual pilots. They lack the colour and spark of individualised markings. They are devoid of the names of the pilot or crewchief, stencilled on canopy rails with purpose and pride in other air forces. The five-digit FV number, or serial number of his Viggen, is painted in numerals less than an inch high on the rear fuselage, too small to be read, even in a sharp photo of the full aircraft. Viggens wear no nicknames, no nose art.

He gets a mission briefing, gets a weather briefing, and files and flight plan. He checks out gear and hardhat from the equipment room. His helmet is individually contoured, but it, too, lacks the flair of individual markings, and while many air arms camouflage their helmets, his is bone white. His flying suit is similar to the standard NATO immersion suit, having watertight rubbers at its extremeties. It differs, however, in that parts of the suit have a series of 'pipes' – which connect to the Viggen's built-in environmental control – through which cold or heated air can be passed to assure an even temperature.

The parachute is already aboard the aircraft. After a walk-around check, a scrupulous one using a memorised checklist, and after climbing aboard with an external ladder (the Viggen has no built-in steps) the pilot will be connected to seat, chute and control system at seven points. Leg restraints are used but their attachments are built into the back of the flying boots rather than being separate garters. 'You don't strap in', says Svensson. 'You spend 10 minutes connecting yourself to it'. It has been noted that the cockpit is spacious, 'ergonometric'. The instrument layout is determined, rather sensibly, by the location of head-up and head-down display units. Almost every switch is in easy reach. A bit odd, at first glance, is the T-shaped control on the right-hand console used to fire the Rb.05 air-to-surface missile.

Getting Underway
An auxiliary power unit (APU) is used to start Viggen's highly praised RM8 engine, even in remote locations such as the established hardstands tucked into trees along road runways. The Viggen's engine can be self-started without an APU but this runs down

Left:
Like the pilot of any aircraft, the Viggen flier is going to make a walk-around check before climbing aboard. With the Viggen, the procedure consists of far more than kicking the tyres. Here, a pilot opens an access panel just aft of the Viggen's nose radar.
Swedish Air Force: Rune Rydh

Top right:
While another pilot watches, a Viggen flier settles into his cockpit. The aircraft has its auxiliary power unit (APU) connected and is ready to start.
Swedish Air Force: Rune Rydh

Above right:
Towbar and APU disconnected, a Viggen is ready to roll.
Swedish Air Force: Rune Rydh

batteries and increases risk during take-off. At ease in his cockpit, canopy closed, the pilot communicates with his ground crew up to the point of APU disconnect, visual clearance and taxi-out, using pre-ordained sign language. An intercom connected to the aircraft via an umbilical was tested extensively in 1982 but is not often used.

The Viggen pilot is, of course, integral now with his ejection seat. A combined parachute and seat harness is used, a barometric interlock releasing the occupant and harness from the seat if he must eject, as can happen, with zero-zero capability, even on the ground. The manual override handle for this sequence is on the left-hand side of the seat pan. If need be, as is *not* possible in the F-4 Phantom for example, the pilot can eject through the canopy.

He has gotten this far, ready to go now, because he has had top-notch support. Viggen requires fewer ground personnel than any other aircraft in its class. With a ground crew of only seven men or women, including five conscripts with limited training, Viggen requires but 10 minutes for refuelling and rearming between fighter missions. The Viggen can remain at high alert for days on end. An external electric source keeps the onboard electronics operational and also provides air conditioning. Viggens on full alert can be aloft in 30 seconds and reach 32,800ft (10,000m) in less than two minutes.

Waved to taxi out, the pilot stays on tower frequency; or, if flying from a roadway, the frequency of a mobile command post. His radio callsign is based on the phonetic alphabet given his wing, plus the two-digit tactical call number painted on the aircraft tail. If he is flying AJ37 37019 of F6 wing which has 19 painted on the tail, his callsign is FILIP 19 (F for Filip being the 6th letter of the 30-letter Swedish alphabet and, hence, used for F6 wing).

Into the Mission

Moving toward runway's end at minimal throttle, the widetrack Viggen main gear provides easy ground handling and the unbroken, birdstrike-resistant windscreen affords superb visibility. The sitting position is high and the excellent view takes in almost a full 360° but for the inevitable rear blind spot. The spaciousness of the ergonometric cockpit and its climate control makes FILIP 19's pilot far more comfortable than, for example, the pilot of an F-4 Phantom whose seat is narrower and whose personal gear has a reputation for being bulky and uncomfortable.

Nosewheel steering using the rudder pedals simplifies taxying in the Viggen. A thumb tab on the upper backside of the control stick increases the turning range of the nosewheel when depressed, but is used only at low speeds so that the airframe is not overstressed during high-speed ground operations. This tab is hinged at the top and flips over to the front side of the stick to become the weapon firing trigger.

The dramatic, thunderous, uncommonly short take-off of the STOL Viggen was introduced at the outset of chapter one. The English language is used for most key communications: 'FILIP 19 is clear for take-off', booms the tower man's (or woman's) voice in the earphones. If a two-plane element is making a

Below:
The Viggen pilot's cockpit is among the world's most advanced. Shown is the cockpit of a second-generation JA37 interceptor. *Saab, via Michael France*

Right:
Preparations are essentially the same for a reconnaissance mission in the camera-equipped Viggen. Here, a pilot prepares for a flight in the photo-taking SF37 craft of F17 wing at Ronneby. *Swedish Air Force: Rune Hedgren*

Below right:
The Viggen's classic take-off roll, on full afterburner. *Swedish Air Force: Rune Rydh*

side-by-side launch together, the pilot of the number two Viggen watches for the slight motion of the main-gear strut on the lead fighter, telling him that lead has released brakes. He opens throttle, applies afterburner, releases brakes, and goes.

Airborne
Take-off. Six seconds elapse from brakes-off to rotation. From a standing start, FILIP 19 blasts forward to 185mph (300km/hr) in those six seconds. Rudder control comes in easily with movement and only a slight pull on the stick breaks the 35,275lb (16,000kg) fighter free of earth's bonds. FILIP 19's pilot tucks in gear and goes off reheat. He climbs at a severe angle over breathtaking Swedish forestland. Viggen's thrust-to-weight ratio, about 0.82 to 1 with burner, is far from superlative but even without reheat he can stand it on its tail if he must, with a trade-off of climb rate for fuel burn.

Wartime missions in Viggen's contingency portfolio – straight up on a collision-course GCI intercept, lo-lo for short-range attack in an intense threat environment, hi-lo-hi for the strike role at greater range against thinner defences – exceed the scope of this account of

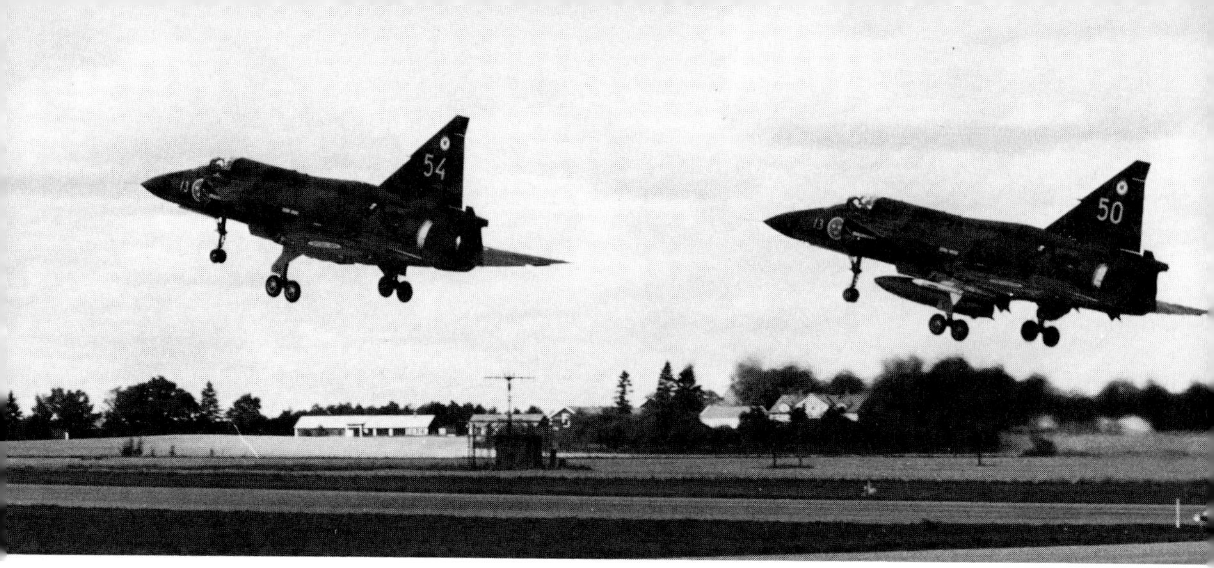

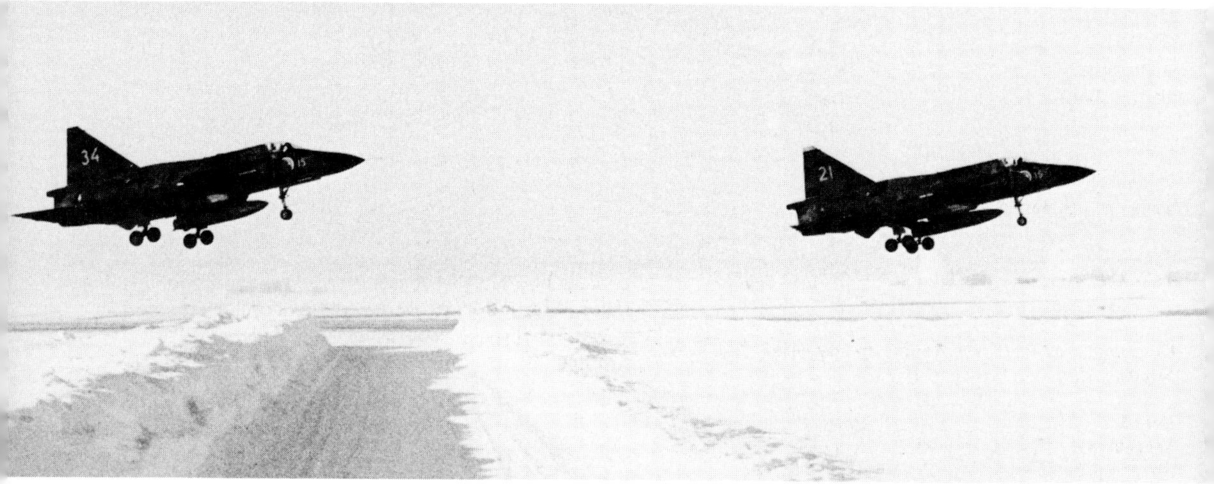

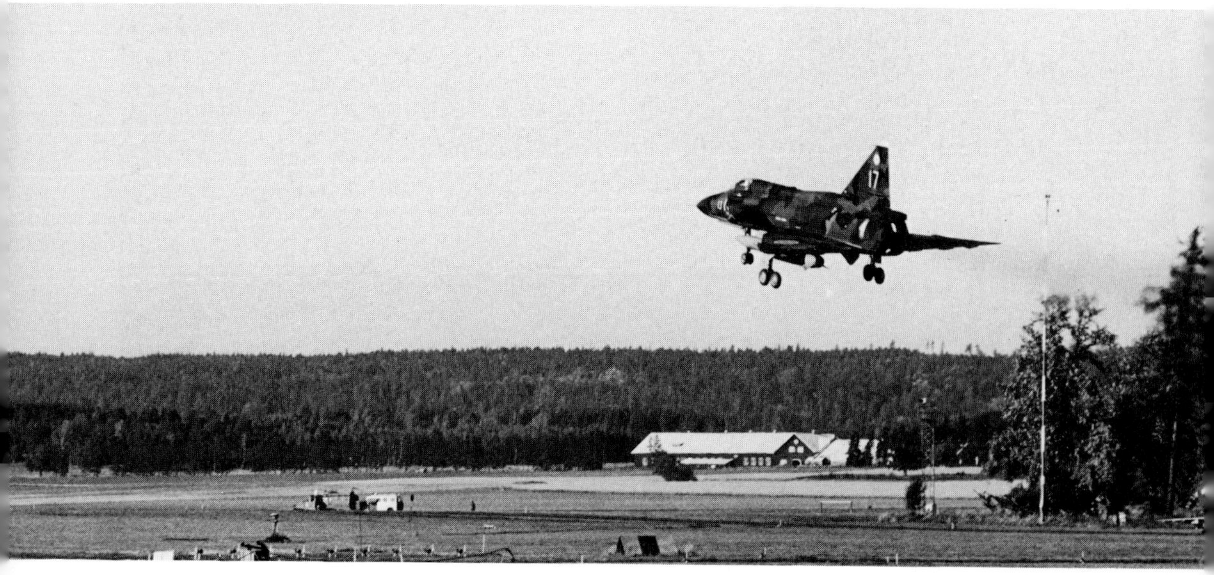

the 'feel' of flying the airplane. Mission profiles are not published. Viggen pilots say that foreign impressions of their mount's 'shortleggedness' – high fuel burn, limited range – are exaggerated. Further, they fail to acknowledge that long range is not essential to Sweden's peculiar defence needs. Ordnance and fuel load naturally dictate not merely combat radius – which, obviously, is also influenced by altitude – but the airplane's speed, manoeuvrability, and survivability in close-quarters combat. A clean Viggen, stripped of centreline fuel tank and slung ordnance, can be thrown into violent air combat manoeuvring (ACM) at high G's with minimal risk to airframe structure or pilot discomfort. In some regimes only the Northrop F-20A Tigershark can turn inside it. It is unclear whether the thrust reverser can be employed in flight but, if it can, presumably the Air Staff would like a MiG-29 pilot to find out the hard way in a clawing ACM duel. For our mission, FILIP 19's pilot merely climbs away at the modest standard cruising speed of 400mph (650km/hr), or Mach 0.52. He can go supersonic whenever he wants and scarcely feel it but, again, with a trade-off in fuel depletion.

Top left:
Take-off. *Swedish Air Force: Rune Rydh*

Centre left:
Take-off, from snow-covered Sweden. *Saab-Scania*

Bottom left:
Climbing away from Norrkoping on take-off.
Swedish Air Force: Rune Rydh

Below:
Viggen aloft, in this case a SH37 model (37302), over Swedish forests. *Saab-Scania*

Some idea of what Viggen can do was given without figures by SAAB's M. Ingemar Olsson at the Paris Air Show on 25 May 1973:

'Although the Viggen has extreme performance capability it is extraordinarily easy to fly all the way out to its limits. This has been confirmed by the Swedish Air Force's first operational squadron as well as by pilots from foreign evaluation teams who have flown Viggen recently.

'The Viggen includes a versatile and very flexible up-to-date electronic system. But in spite of that – by means of an advanced system integration and biotechnological development – the pilot has to devote more and more of his capacity to what is happening outside the cockpit.

'The pilot shall not have any need to look down into the cockpit during an attack or air fight. Consequently, we are now incorporating things like 'extended-view HUD', quick-access modes for guns and dogfighting missiles, automatic radar lock-on and tracking and an aircraft energy-management guidance system.'

If he is flying an attack mission against a high-threat target – say, an invasion fleet offshore in the Baltic, confronting him with air cover, radar-directed guns and surface-to-air missiles (SAMs) – the AJ37 pilot of FILIP 19 carries pod-mounted active jamming and passive decoy dispensers. The airborne use of expendable decoys – chaff to foul enemy monopulse radar, infra-red flares to draw off heat-seeking missiles – was much-publicised in the Falklands conflict and resulted in an array of such countermeasures being developed by Philips Elektronikindustrier, the Swedish supplier which manufactures more chaff and flare dispensers than any maker outside the United States. FILIP 19's pilot is also attentive to his radar warning receiver (RWR) although he may wish that Stockholm had forged ahead with plans, pondered in the mid-1970s, to acquire the Grumman E-2C Hawkeye airborne early warning (AEW) platform to help him identify and evade threats. Those plans – weakness in AEW, remember, was a negative lesson of the Falklands fighting – fell to the budget-cutters' axe. But FILIP 19 still has its countermeasures and the pilot has been coached in high-speed 'jinking' to avoid a SAM when it is fired at him.

Little is known of the doctrine and tactics developed by *Flygvapnet*. It must be assumed that on this mission, top cover (MiGCAP) is provided by JA37 fighters from one base while the attack mission is pressed home by FILIP 19 and wingmen from another. Relying heavily on data-link information presented to, and displayed by, his digital computer system, the pilot uses his head-down display (HDD) for target information and ranging as he dives to the attack. He may engage and launch ordnance like the Rb.05

Into the mission: Viggens airborne. *Saab-Scania*

anti-shipping missile without ever coming within eyesight of the offshore fleet.

Or he may press in for a point-blank assault. FILIP 19's pilot applies full power and burner now, rechecks his weapons settings, arming switch, airspeed and altitude. He rolls in. Peering, straining, leaning forward in the harness, he is alert to the target and its defences. He may close in to blast away with pod-mounted 30mm Aden cannon. He will be studious and serious, but the adrenalin is churning. The lead-computing imagery on the head-up display (HUD) gives him ranging and other information as he acquires the target visually, uses decoys to foul its defences, and comes within firing range. Because the cannon is mounted in a ventral position, there is no danger of gunflash glare and the weapon has surprisingly little effect on the forward speed or stability of the Viggen itself. Shells striking home – against, say, a destroyer turning in heavy sea – do *not* walk across the target the way they do in the movies. To the onwatcher it appears that every shell in each burst arrives at the same instant.

Of course, no Viggen has yet seen combat. Most flights are of relatively short duration. Most pilots, as noted earlier, log comparatively few flying hours in the course of a year. The ground simulator for the Viggen is so much like the real airplane that it is said to provide around 80% of the realism acquired in actual flight. The typical Viggen flight may be little more than a routine hop a couple of hundred kilometres away from base, with return to the same base.

Like just about every new fighter design in recent years, the Viggen is sometimes called the 'missile with a man in it'. Its automated systems are cited as evidence that the fighter pilot, like the dinosaur, has become obsolete. With its data-link computer systems and 'hands off' capabilities, the Viggen *can* quite literally be flown from take-off to touchdown without the pilot touching his controls. But *Flygvapnet* acknowledges that no real substitute exists, yet, for the man trained and motivated to fly and fight: in the end, FILIP 19's pilot has fully as much control over his destiny as the pilot of a World War 1 pursuit ship. In an air battle, or solely to test the airframe to its limits, his final resort is the toolery found in all flying machines – stick, rudder and throttle.

In contrast to the fuel warning light on the Phantom, which is supposed to signal only 500lb of fuel remaining and is accurate only to within 200lb – the cause of some gasping, bone-dry arrivals from South-East Asia combat missions – the Viggen's 'bingo' light warns the pilot when he has 24% of fuel capacity remaining (the exact figure of the aircraft's fuel capacity is undisclosed).

It must be repeated that Viggen is not in any event a long-range machine, that Sweden makes no use of air-to-air refuelling, and that fuel starvation is not generally seen as a problem because of the short duration of missions flown.

A tactical ILS approach is standard *Flygvapnet* procedure. The approach to base is flown manually, but with auto-throttle. Stability augmentation is

Above:
Viggen pilots after a successful mission, ebullient with the shared sense of purpose unique to men who fly and fight. Pilots return to base with virtually no fatique, thanks to the controlled environment of Viggen's roomy and comfortable cabin. During an actual mission, with head-up and head-down displays to assist them, pilots have relatively little need to refer to, or jot notes on, the clipboards appended to their kneepads.
Swedish Air Force: Rune Hedgren

Left:
Drink for a Viggen's pilot, after a mission, contains undisclosed liquid. Medicinal rum is hardly needed.
Swedish Air Force: John Forsell

available but not needed for a craft which behaves well at low airspeeds even in turbulence. A 12° angle of attack is reached through appliction of autothrottle for a normal letdown, where full STOL capability is not needed, though a 15° attitude can be employed. Glideslope is held by stick movement and the auto-throttle holds speed and altitude, automatically making adjustments where necessary to hold the aircraft on its high-angle approach. Because, as noted earlier, the Viggen makes a no-flare, carrier-type landing, the pilot simply holds the aircraft at its high-angle all the way to the ground, not needing to make the major last-minute adjustments required at runway's end for more demanding and less forgiving machines. It is worth noting that experienced pilots who have a special love for Draken which Viggen fails to evoke, find the Viggen far easier to land.

8 The Viggens that Weren't

Above:
The classic lines of Viggen, as seen from straight ahead in flight – one of many SAAB publicity photos which were in circulation at the time when Viggen was a potential candidate for foreign sales. *Saab-Scania*

More than the Viggens which are, the imagination is stirred by the Viggens which are not. The Viggen did *not* become the standard Royal Air Force fighter of the 1970s. It never flew with the rising-sun emblem of Japan, never guarded a Norwegian fjord, never wore NATO colours. Austria and India were among nations which wanted the Viggen, but didn't get it. Politics, fate, and a mind-boggling array of unpredictable events seem, at every turn, to have dashed hopes for foreign sales of Sweden's principal fighter. Yet between 1967 and 1977 the Viggen came close to being adopted by numerous air arms. The Viggens that never flew outside Sweden are as much a part of the Viggen story as those which did. While arms-transfer policy, political manoeuvre and national pride are beyond the scope of this work, the following tour d'horizon is an essential part of the story of a great warplane that might have been more widely employed.

Above:
Replete with cherry blossom on its centreline fuel tank, a plastic model made from a 1/72nd scale Heller kit represents the fictitious Mitsubishi/SAAB 37-J Viggen (serial number 37-8333) stationed at Chitose Air Base in Northern Japan. Japan looked seriously at Viggen, but did not buy. *Anders Nylen*

Great Britain

Might the Viggen have served as the backbone of the RAF's fighter force? It was never especially likely. At best, the idea of a British Viggen was a pipe-dream for a brief time, entertained by a few minds in London and Stockholm.

There has always been a 'British connection' with the Viggen project. In March 1968 SAAB placed an order valued at £150,000 with the Hemel Hempstead, Hertfordshire firm Avia for bellows-sealed flexible joints in nimonic and stainless steel for the high-temperature air-ducting system on the AJ37. A year later an order for 200 sets of fire detection equipment for the AJ37 went to Gravinor (Colnbrook) Ltd in Cambridge. This order was for 'firewire', temperature-sensitive elements which, mounted around engine and control units, give a cockpit warning in the event of fire or excessive temperature rise; the same British-made elements were used for fire warning in the SAAB J32 Lansen, J35 Draken and Sk.60 (SAAB 105) trainer.

Even more important, to give the Viggen an Anglic touch, the leading edge of the canard foreplane of the AJ37 has Aeroweb aluminum honeycomb bonded with BSL 308 high-strength adhesive, both products of CIBA (ARL) of Cambridge. The company has become a regular supplier to SAAB for honeycomb materials so vital to Viggen design.

In August 1967 a discerning *Financial Times* correspondent in Stockholm reported discussions in the Swedish capital between un-named British visitors and Air Staff officers over the Viggen as an alternative to the recently-cancelled Anglo-French Variable Geometry Fighter (AFVG). When Defence Minister Dennis Healey had announced France's withdrawal from the ill-fated AFVG project the previous month, he had specifically not ruled out the possibility of a British purchase abroad. Swedish Defence Ministry officials were quick to say that they were always open to co-operation with Britain. SAAB, not surprisingly, had no comment.

On Thursday 17 August 1967, in one of the quickest sales trips ever undertaken, a Swedish mission arrived in London for talks with the Ministry of Defence about Viggen as a British AFVG replacement. The head of this delegation was no lesser personage than Gen Lars Brising, earlier the technical director at SAAB and now Director-General of Swedish Air Force Engineering and over-all co-ordinator of the Viggen programme. A long Friday, 18 August, was spent in discussions with MoD. Who said what? Was this a courtesy visit or was it a serious undertaking? The record shows only that the Swedish delegation promptly returned to Stockholm on Saturday. Had they been 'shot down'? No one who will talk about it knows; no one who knows will talk.

In September 1967 the usually well-informed *Flight International* reported that proposals to the RAF for an anglicised Viggen were being carefully studied by the Ministry of Technology. What, in the meanwhile, had happened to the Ministry of Defence? *Flight* reported that a British engine would be used in the RAF Viggen, but also said that chances for the proposal seemed remote.

Roy Braybrook, aviation writer and longtime member of the Kingston-upon-Thames fighter team which has developed great aircraft from Hurricane to Harrier, remembers that the idea of an RAF Viggen was still under discussion in 1968-69: 'The Viggen matched RAF requirements for a strike aircraft.' Braybrook went to Sweden to brief Stockholm's Air Staff on what must have seemed, then, only a passing curiosity, a very peculiar British plane called the Harrier. 'We saw Harrier as a lightweight complement to the Viggen and people were talking about a hi-lo mix', possibly an arrangement in which both air forces would employ both Harrier and Viggen. Braybrook does not remember how serious his British colleagues were about purchasing the Viggen, nor his Swedish hosts about selling it. 'We were certainly in deadly earnest about selling them the Harrier.' Sweden, of course, never purchased the British V/STOL fighter, which was later to triumph in the Falklands – although the STOL requirements for the Viggen show an uncommon appreciation of the vulnerability of fixed runways – and, as it turned out, there never was a British Viggen.

Says aviation writer Bill Gunston, 'There never was a strong lobby for the Viggen in Great Britain. Some people saw one of the aircraft's great faults in its very high fuel burn'. Roger Bailman, an MoD official, remembers that 'some people, at least, were very keen

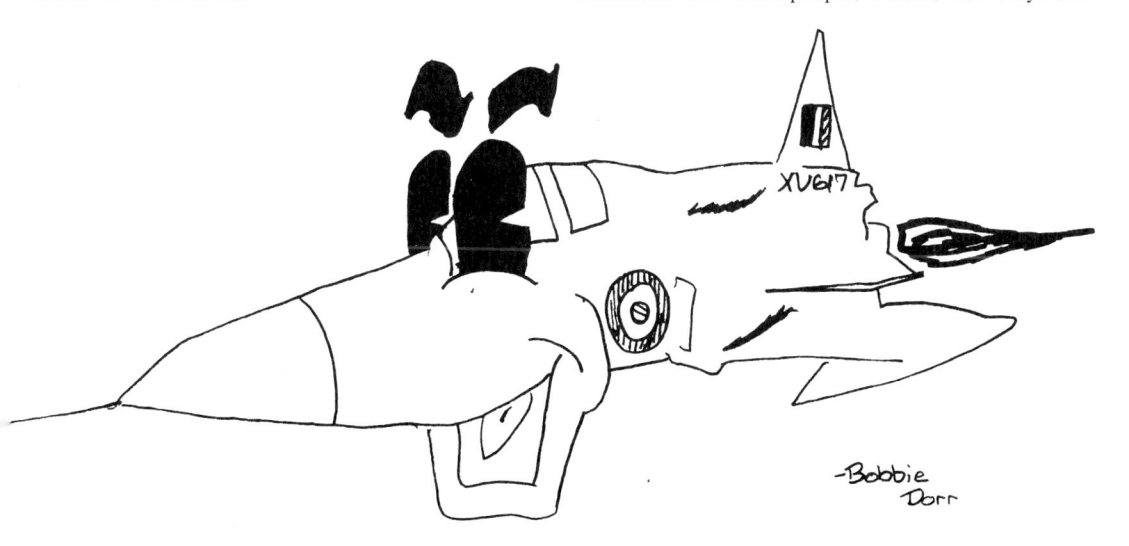

Below:
Might the Viggen have become the standard fighter of the Royal Air Force? Fanciful caricature by author's 13-year-old son shows what an RAF Viggen might have looked like. *Bobbie Dorr*

99

on Viggen. They liked the radar and the weapons-system intergration. They thought a longer-range version would have real potential for the RAF.' But it was not to be.

Norway
In 1967-68 Norway was interested in Sweden's STRIL 60 air-defence system and was eyeing development of the Viggen closely. A kind of friendly antagonism has always existed between the two neighbours on the Scandinavian peninsula, and their differences at times exceed the mere difference between neutrality and NATO membership. Still, discussions were held in Oslo and Stockholm with a view toward a standardised air-defence network, with a standard fighter aircraft to be used by both nations. 'Co-production would have been an essential part of any such arrangement', says Norwegian diplomat Peter Ettesvold. The discussions did not proceed far, and the Northrop F-5A ultimately complemented Oslo's fleet of Lockheed F-104s. Again, it was not to be.

Australia
In June 1972 SAAB formed a subsidiary in Australia with the intent of working toward a Viggen order from the Royal Australian Air Force (RAAF). Canberra was at that time contemplating a replacement for its Mirage aircraft.

Sweden informed Australia that no limitations of political origin would be placed upon any Viggen sold to the RAAF. As it turned out Australia postponed a decision on a new fighter, gaining more fatigue life from its Mirages by putting them back on to pure interception duties and deleting an attack role. Ultimately, the RAAF resolved its controversial purchase of the General Dynamics F-111C and has also become a purchaser of the McDonnell F/A-18A Hornet. Australian pilots reportedly liked the Viggen but planners were concerned about the distance between supplier and user.

'The Deal of the Century'
Le marche du siecle, the Belgian-created term for the 'deal of the century' under which NATO nations would order 348 fighter air-frames to replace the Lockheed F-104 Starfighter in the late 1970s was, from the beginning, an exaggeration. It was assumed that the winner of this fighter competition would also be ordered by the US Navy, Canada, Australia and other nations (many of which, in the event, settled on the F/A-18A Hornet instead). It was assumed that the fighter chosen to guard NATO might also be picked by Japan (which turned out to prefer the F-15 Eagle). In the end, the 'deal', which involved co-production and created tens of thousands of jobs, was a major transaction all right, but smaller than many others – including, for example, Sweden's commitment to the Viggen for its own use.

Competitors for the deal included the French Mirage F.1E, the proposed Northrop F-18L land-based variant of the Hornet, the General Dynamics F-16 Fighting Falcon, and Viggen. The head of SAAB, Curt Mileikowsky, is quoted as saying that Sweden had only an 8% chance of winning the massive order for its principal fighter but that 'it was a chance worth taking'. He felt Denmark and Norway might be attracted by mutual production agreements among close neighbours. SAAB was prepared to offer generous 'offset' arrangements and had gotten its price down to $9million (£3.2million) per Viggen airframe, an almost unbelievable accomplishment with such a sophisticated machine; but in that pre-inflation era the F-16, backed by American sales prowess and political might, was priced at $6.9million (£2.46million). The Pentagon's candidate also had another powerful selling point: the Americans had already selected the F-16 for their forces. The F-16 was to become one of the most effective and widely used fighters in the world and it was the brunt of an inevitability which could not be obstructed.

In the earlier consideration of the Viggen by Great Britain, and by Norway, as with the 'deal of the century', there had been a powerful argument against Viggen. Although Sweden keeps its aircraft and guns pointed east, it is a neutral nation. It has no commitment to mutual defence against an assault by the Soviet Union and the Warsaw Pact. Some critics believed that, during World War 2, Stockholm had been 'less neutral' toward whichever side was winning at the time – accommodating German passage to occupied Norway in 1941, allowing American aircraft to operate in its northland in 1944. If there were a war in Europe, could NATO partners like Holland, Belgium, even Norway, rely upon support, parts and overall assistance from a friendly but non-aligned state? Says Carl-Olov Munkberg, a Swedish aviation editor, 'They would not have accepted our airplane unless we had joined the NATO alliance, which was never a realistic possibility for us.'

The SAAB 37E Eurofighter, the second-generation Viggen variant thrown into the competition and heavily promoted in the early 1970s, still had characteristics which were ideal for the Swedish defence situation but were anathema to NATO. It remained a highly sophisticated machine and if it demanded relatively little maintenance for such a degree of sophistication, it still required more than simpler fighters. It remained a relatively short-range craft with high fuel consumption. It was a forgiving airplane, unlike Draken, but not easy to master. How would transition training be conducted? Training squadrons for the F-16 were already flying in the great American deserts, and a training syllabus was being worked up for pilots and ground crews from nations as disparate as Pakistan and South Korea.

Once again, it was not to be. The NATO purchase, if not history's largest financial transaction, went to the formidable F-16. There would be, could be, no NATO

Above:
The winner of the 'arms deal of the century', the fighter chosen to replace nearly 400 F-104s in NATO inventory, was not the Viggen but the General Dynamics F-16 Fighting Falcon (a two-seat F-16B is seen here). Later Sweden would consider the F-16 among foreign possibilities as a later-generation replacement for the Viggen in its own forces, but would decide upon the indigenous JAS39 Gripen instead. *Michael France*

Viggen. Researcher P. A. Sammons believes that Viggen never really had a serious chance, – not even the 8% chance. 'It is a fine, fine fighting machine but the circumstances were not right.' Again, once again, it was not to be.

Japan

In July 1973 Japan was considering licence manufacture of the Viggen. A Country for practical purposes non-aligned (though it is signatory to a Mutual Defence Treaty with the US) Japan had needs not dissimilar to Sweden's and might have adopted a Viggen with co-production by the Mitsubishi firm in Nagoya. 'Might have', repeats William T. Randol, an authority on Tokyo's defence posture. 'But there were all sorts of problems and simple distance, plus the "special relationship" with the US, were among them.'

Japan selected the F-15 Eagle as its fourth generation interceptor in December 1976 after a lengthy evaluation which initially considered no fewer than 13 candidates, including Viggen. The seven still in contention by 1975 (to supplant, but not fully replace) the F-4EJ Phantom already built under licence by Mitsubishi were Tornado, Mirage F.1E, the F-14 Tomcat, F-15, F-16, F-18 and Viggen. By January 1976, Sweden's pride had been deleted and three contenders remained: F-14, F-15, F-16.

In 1978 in Tokyo, reflections on what might have been were offered by Col Toshio Hayata, commander of the Japanese Air Self-Defence Force (JASDF) base at Chitose in the far north where, as they do in northerly Sweden, fighters set forth to intercept and escort Soviet intruders over international waters. Hayata had been to Stockholm but would not say when or why. 'We had language problems. We had different concepts. We tried to explain that a patrol from Hokkaido (the northernmost Japanese island) required greater mission endurance than a similar patrol into the Baltic. We thought the Viggen was an excellent "pilot-oriented" machine and were much impressed by it.' Swedish modeller Hans Percy is so intrigued by what might have happened that he constructed a finely detailed model of a ficticious Japanese Viggen, licence-built Mitsubishi/SAAB AJ37-J serial number 37-8333 in the markings of the Chitose-based squadron. Percy's superb craftsmanship is all that exists, now, of another foreign Viggen which, again, was not to be.

Above:
Aircraft that never existed have rarely been portrayed so well as in this photo of a Swedish Air Force McDonnell F/A-18A Hornet. The F/A-18A came along too late to be a serious competitor with Viggen for the Nato 'deal of the century', but was among the foreign designs studied by Swedish officers as an eventual replacement for the Viggen, before they settled on the JAS39 Gripen design. *McDonnell Douglas*

Austria

In some ways Vienna is an unlikely capital in which to sell airplanes. Budgetary constraints are so great that the very existence of the *Osterreichische Luftstreitkrafte* (OLK), or Austrian Air Force, remains perenially in doubt. But the Swedish connection is there.

At one time, with Viggen looked at, considered and evaluated by a dozen nations, it looked as if the only export order for the aircraft might come from Austria, itself neutral, and thus not subject to the political concerns of some other potential users. Austria seemed a 'safe' customer. It was one of the few successful users of SAAB export aircraft, including the SAAB 105 (Sk.60) trainer and the J29 'Flying Barrel' fighter of the 1950s. It had evaluated the Israeli Kfir fighter and found it wanting. The 'strong export potential' which *Aviation Week* had predicted for the Viggen back in May 1967 clearly wasn't there, but *surely*, thought Stockholm's planners, *surely* at least Austria could be depended upon!

It could not. Indeed, since the 1970s the budgetary restraints on the OLK have increased so greatly that, as this book went to press, Vienna was wondering not if it could afford Viggens but whether, in a drastic economy measure, it might be able to pay for earlier-generation J35D *Drakens*.

India

Viggen was a serious contender for a fighter order from India in 1977-78 when the Carter administration, following a short-lived arms transfer policy which many also saw as short-sighted, used its hold on the American design of the engines to scotch the deal. It has remained unclear whether Viggen might have been the Indian choice but for US resistance, which, during the same period, killed a sale of the A-7 Corsair II to Pakistan. The Indian Defence Minister, Jagjivan Ram, announced in October 1978 that in preference to the French Mirage F.1 and the Viggen, India would order the Sepecat Jaguar as its next tactical fighter.

9 'It's a Combat Situation Out There...'

The weapons systems officer (WSO) cramped in the back seat of the West German RF-4E Phantom cranked his head around, peered to the limit of the Phantom's rearward visibility, and blinked. 'There they were', he says. 'They had come up on us so fast, we didn't appreciate that they were about to form on our wing. They could as easily have poured 30mm cannon fire up our tailpipes.' Two JA37 *Jaktviggen* interceptors had nailed the Phantom over international waters, high above the Baltic. 'If they had been hostile, we would have been meat on the table for them', says the German WSO. The JA37s flew escort off the Phantom's wingtips, German and Swedish airmen exhanging friendly hand signals.

Sweden is neutral. It has no alliances, claims no enemies. Yet the subconscious thinking of Viggen pilots is easily understood. A caption on the back of a photo of a US Air Force RC-135 reconnaissance aircraft – similarly intercepted by Viggens over the Baltic – reads: 'Sometimes the aircraft we intercept are flown by *friends* . . . '

The rakish Viggen fighter which guards Sweden today never won a single foreign sales order, never flew in the colours of the NATO countries, India, Japan, or any of the other nations once interested in it, and has not been in a shooting conflict. Sweden itself, except when supporting United Nations operations in the Congo 1961-65 has not been a participant in a war since the beginning of the last century. Yet the young pilots who fly Viggens out over the contested Baltic come into close contact with Soviet intruders with growing frequency, and are ready to use guns instead of cameras if necessary. In the meanwhile their cameras have brought back some of the clearest photos of Soviet aircraft seen in the West. The Swedish Air Force is retrofitting its JA37s with a small camera,

Below:
Viggen pilot, getting ready to go. This craft from F17 wing at Ronneby, only a few miles from where a Russian submarine came ashore in October 1981, may soon be heading out for the Baltic Sea on a mission under realistic combat conditions. *Swedish Air Force: Rune Hedgren*

103

designed to be operated by the left hand, to permit pilots performing intercepts on intruders at the fringes of Swedish airspace to get better photographs of the aircraft they have intercepted. The choice of a left-handed camera was made to permit the pilot to take photos without removing his right hand from Viggen's controls.

It is only a bit of an exaggeration: when one Viggen pilot returned from a defensive patrol in the Baltic, he shook his head seriously, adopted a serious expression, and said, 'It's a combat situation out there.'

Viggen pilots remain especially wary toward Soviet aircraft, including the Tupolev Tu-26M ('Backfire') bomber operated by Russian naval units in the region. Before development of the STRIL 60 air defence system, an intruder might easily have slipped into Swedish airspace and, if bent on harm, could have inflicted it. This was demonstrated in 1947 when a propeller-driven Lavochkin LA-9 fighter flew more than 100 miles (160km) into Sweden before cracking up at Tullinge, not far from a major fighter base, without ever being intercepted. In that incident, the Lavochkin, was flown by a defector and no harm was done. But other incidents, part of history yet fresh in memory, demonstrate why today's Viggen force maintains a high state of readiness.

The Cold War is not yet spoken of in the past tense. Too fresh in memory is 'Friday the Thirteenth', 13 June 1952, when a Douglas DC-3 transport, designated Tp.79 by *Flygvapnet* (serial 79001) went out into the Baltic carrying its flight crew plus warrant-officer signals trainees, eight men in all, and never came back. 79001 was to be the first aircraft flying from Swedish territory to be shot down by hostile action since man learned how to fly.

The DC-3 had apparently been engaged by Soviet MiG-15 fighters and shot out of the sky. Two days later, on Sunday 15 June, the Swedish destroyer *Sundsvall*, 35 miles (55km) northeast of the island of Gotska Sandon, north of Gotland, found a rubber dinghy from the DC-3 burnt and blackened by cannon fire. The plane and its crew were never found, but this grave incident was not yet finished.

On Monday 16 June, in early morning, one of two Swedish Consolidated PBY-5A Catalina air-sea rescue amphibians, designated Tp.47 and serving with F2

Left:

A Viggen pilot climbing aboard his mount, at F15 wing at Soderhamn. *Swedish Air Force: John Forsell*

Above:

Going aloft. The Russians may be waiting.
Swedish Air Force: John Forsell

wing at Hagernas (serial 47002), was over the Baltic searching for survivors from the DC-3 when it was fired upon without warning and forced down on the sea by two Russian MiG-15s. This happened about 22 nautical miles (40km) southeast of the Finnish lighthouse island Bogskar, south of the Aland archipelago. The Catalina, unarmed, with seven men aboard, had been victim of an unprovoked assault. Its pilot had one arm severely injured from a 'near miss' by a MiG cannon shell before he made a forced landing at sea which saved the crew members' lives. The Catalina itself was so badly damaged that it sank while the crew was being rescued.

Swedish pilots remember that diplomatic protest, charge and counter-charge, had little effect – and that both aircraft shot down by the Soviets had been unarmed and utterly defenceless.

'This is serious business', says a Viggen pilot. In recent years, with the Soviets seemingly unable to mask their intentions, much of the publicity of the potential for trouble in the Baltic has been devoted to submarine activity. The SH37 maritime reconnaissance Viggen can, when assigned, operate on anti-submarine warfare (ASW) missions in conjunction with the Swedish Navy's Vertol 107, or Hkp.4, ASW helicopters. Those helicopters have dropped live depth charges on several intruding submarines in the 1980s with unknown result.

Any Swedish pilot smug in the comfort that his nation hasn't been in a declared shooting war since 1814 need only remember that in October 1981 it could have been. That's when the 'Whiskey on the Rocks' incident hit Sweden like a storm. When the Soviet Union's gunmetal-grey 'Whiskey' class submarine No 137 ran aground on a reef in a restricted military zone only nine miles from the Swedish naval base at Karlskrona, the brass hats in Stockholm had to face an unnerving question: was their ASW capability good enough to prevent Russian U-boats from prowling their waters at will?

For a time, the situation was tense indeed. Some people imagined Swedish troops boarding Captain Pyotr Gushin's stranded vessel and hauling its 50-man crew ashore to be interrogated. SH37 Viggen air patrols discovered a Soviet armada standing off in the Baltic, perhaps contemplating moving in to take back the diesel-powered submarine by force. Prime Minister Thorbjorn Falldin revealed to his constituents that torpedoes on the sub 'most likely carried nuclear warheads, according to our investigations.'

In the end the *'Whiskey'* incident was resolved without conflict. Swedish tugboats freed the stranded sub and allowed it to depart. The incident was a severe embarrassment to Moscow – an example of Soviet behaviour later compared, by some, to the August 1983 incident when Russian fighters shot down an innocent Korean Air Lines Boeing 747 jetliner, killing all 269 aboard.

In June 1982, the Swedish Navy's Vertol helicopters had several encounters with a 'foreign submarine' off the northern city of Umea. Vertol crew members, interviewed at that time, were exhausted from long hours of flying and tensed-up with frustration, especially when they believed they had hit a submarine with their ordnance but had no tangible evidence to prove it.

No one seriously believes that a shooting war is going to erupt in the Baltic tomorrow morning, or that either Sweden or the Soviet Union will allow it to happen. The point, of course, is that the SAAB 37 Viggen is far more than a rakish, attractive machine built to inspire young plastic modellers. A serious attitude toward armed defence of the nation's neutrality lay beneath every decision made in the Viggen programme, and behind Sweden's determination to go ahead with the costly and expensive JAS39 Gripen fighter of the future.

At times the price of vigilance is high. Two AJ37 Viggens from F6 wing at Karlsborg collided in mid-air on an operational mission on 5 September 1983 while undergoing air combat training over Lake Vanern. Both pilots, Lt Kjell-Arne Perman and Warrant Officer

Above:
SF37 photo-reconnaissance Viggen over the Baltic.
Saab-Scania

Stefan Lindblad, were killed. In its overall operational history, Viggen has had a more favourable record than most contemporary fighter types – but the pilot always climbs into his cockpit knowing that this can happen.

One of the most worrisome of recent incidents occurred when Viggens were fired upon and the pilots didn't even know it at the time.

In addition to confronting Soviet and other aircraft over the Baltic, Viggens are always encountering foreign warships. SH37s have come upon missile destroyers of the 'Udaloj' class, 'Kanin' class destroyers, 'Golf' class submarines and a variety of other vessels which typify the current Soviet naval build-up. The Swedish Air Force's public affairs officer, John Charleville, told reporters in March 1984 that 'This is happening frequently. We see the routine surveillance of these vessels as essential to our defence.'

In June 1983 two SH37 Viggens from F17 wing at Ronneby were narrowly missed by Soviet surface-to-air missiles (SAMs) fired over the south Baltic. The incident, not disclosed until a year later, took place when the SH37s detected a group of Warsaw Pact warships. The pair of Viggens carried out a low-level camera pass. When the film was developed, it was seen that two missiles had been fired as the Viggens overflew. The pilots had not known about it at the time.

The Swedish Defence staff decided that the missiles

The watchful eye of the Viggen, sometimes the SH37 maritime reconnaissance version, sometimes the SF37 photo ship, has caught all manner of ships and aircraft operating in the Baltic sea. These include:

Top right:
The MiG-23 fighter.

Centre right:
Soviet 'Reductor' class intelligence-gathering vessel.

Bottom right:
This West German Type A52 intelligence vessel.

had not been fired at the Viggens. The incident occurred in international waters and Stockholm decided to take no diplomatic action. But again, the seriousness of Viggen's mission was underlined.

Fighter Base

Ronneby lies in tall birch forests in the grey, wet southeast corner of Sweden, near Karlskrona where that Soviet submarine came aground in October 1981, near the coast where 72% of Sweden's population dwells. During a visit to F17 wing at Ronneby, which

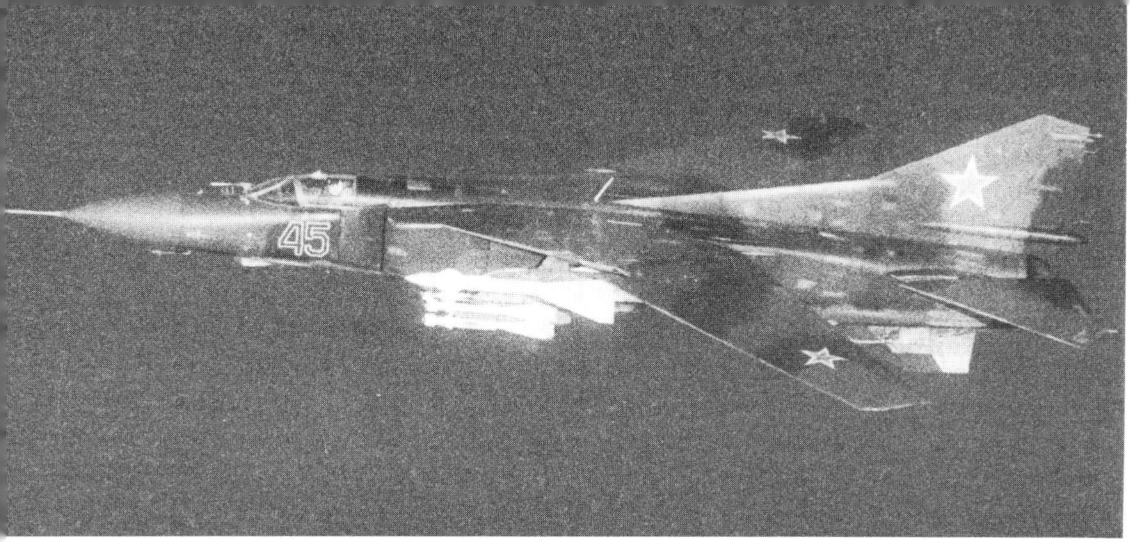
45

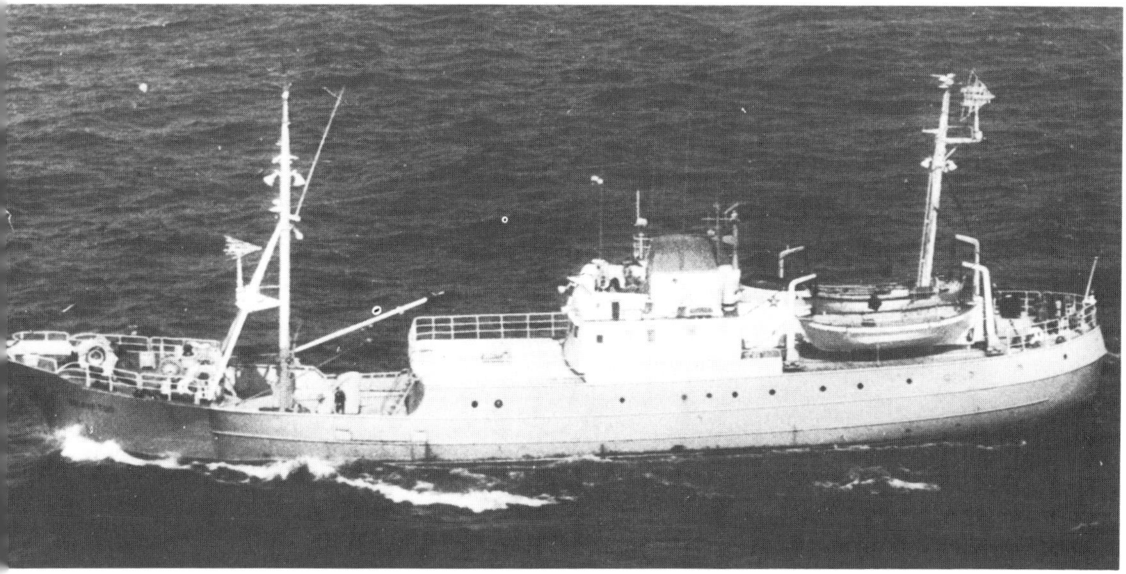

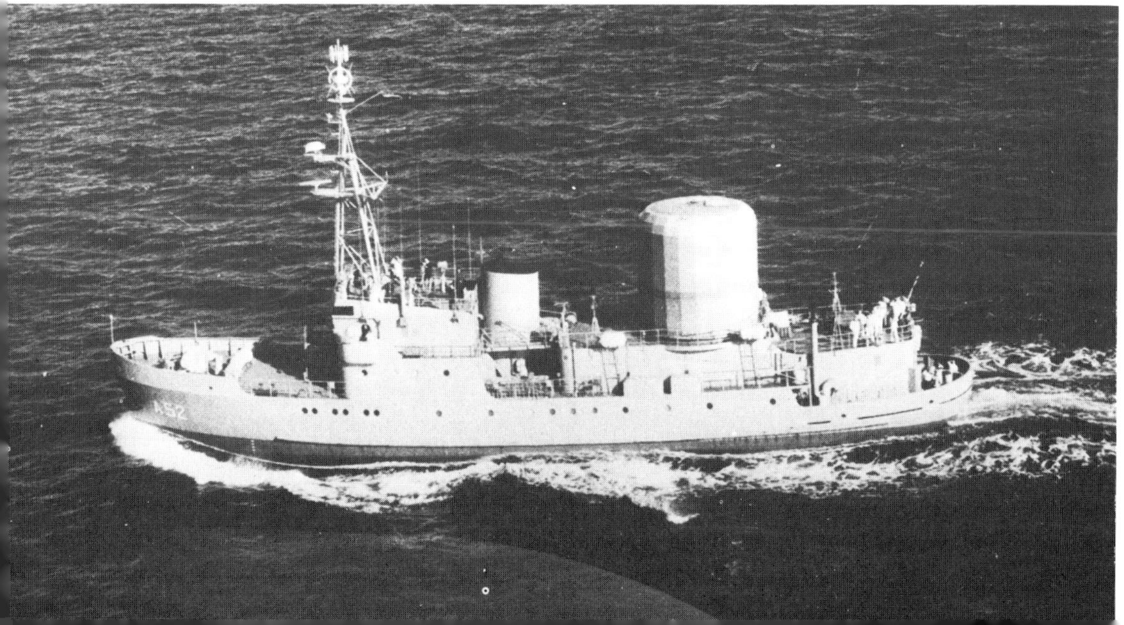
A52

Above:
A Soviet missile destroyer of the 'Udaloj' class.

Below:
An East German Navy corvette.

also lies within easy strike range of Soviet air and naval units confronting Sweden across the Baltic, a serious and sombre mood was evident among Viggen pilots. The O Club is modest. Revelry, drinking, flying with one's hands at a smoke-filled bar . . . these customs seem less entrenched here than at fighter bases in some other nations. The town of Ronneby is so small, it has not a single eating or drinking establishment open past 6pm. Viggen pilots here are not prone to noise-making or high visibility, but they call themselves the best prepared in the *Flygvapen* for the crucible of air-to-air combat. They had better be. History has a relentless way of posing ever-increasing challenges.

Simple numbers tell the story. An air force which had 50 squadrons in 1950 will shortly be reduced to 11. Contacts with intruding aircraft have increased ten-fold over the past decade. An official study indicates that in the event of a low-level, terrain-masked attack, F17 would have precisely 60 seconds to get its Viggens into the air. *Flygvapen's* dispersal plan, known as the Base 90 concept, is intended to protect those Viggens by scattering them at satellite bases using secondary roads as take-off strips, a newly expanded version of the reliance on roadway bases held in mind ever since STOL capability was made a sine qua non for Viggen. Base 90 aims at survival but the conclusion from the numbers is inescapable: with the cost of modern aircraft soaring, with fewer men and fewer planes to face a growing threat, F17 and other wings must have the best airframes money can buy and the finest pilots devotion can develop.

In F17's Red *divisionen*, or squadron, charged with the air combat role, pilots swear by the JA37 Jaktviggen. 'We are ready with this machine', says an F17 safety officer and JA37 pilot. 'We are trained, armed and prepared. The JA37 can take on anything in the world and win.'

Fight and win. That's what Viggen is for. That is the purpose to which young fighter pilots in Viggen cockpits sharpen their skills and ready their wits. No one wants it to happen but, if it does, these men may fly and fight in a wholly different battle arena than the one which existed in the 1950s when Erik Bratt and a few others first saw Viggen as a vision. In Today's world, men may have to fly and fight in a chemical or nuclear battle environment amid the perils of electronic warfare, perhaps even with the certainty that after the mission there may be no home to return to. A world-class fighter aircraft – Viggen – with a few very good men flying it, faces challenges as never before. Says another F17 Viggen jock, 'We can do what must be done!'.

Beyond the horizon lies JAS39 Gripen (Griffin), the fighter of the next century, given the go-ahead only as recently as a Parliamentary decision of 4 June 1983. Five service-development prototypes of Gripen are now taking shape at SAAB's Linkoping plant at a remarkably economical $8million (£5.5million) per airframe. Two hundred and forty may be in service by the year 2005, first in ground attack, later in interceptor, variants. Chosen instead of F-16, F/A-18A and other foreign candidates, resurrected from the ashes of the defunct B3LA program, albeit powered by a General Electric F404J engine developing 18,000lb (8164 kg) static thrust, Gripen will follow Tunnan, Lansen, Draken and Viggen in upholding a costly but

Below:
A Soviet 'Golf' class submarine.

essential tradition – Swedish airplanes for Sweden, designed and manufactured at home, at the fore of technology. Gripen will dispense with the need for an auxiliary power unit for starting and a thrust reverser for STOL landings. It will have a composite wing, a fly-by-wire control system, a holographic head-up display (HUD) with an unprecedented field of view greater than 20°. It will be, in every respect, as advanced as any fighter now extant or contemplated.

Like Viggen before it, Gripen seems certain to traverse a minefield of practical, fiscal and political debate before becoming finished metal and reaching the hands of pilots who will need it, in outlying years, if they are to be able to fight and win.

For now, for years to come – indeed, in production until at least 1988 – Viggen provides that capability. Fight and win. To the fighter pilot, no accolade to an aircraft could be more important.

Above:
A US Air Force RC-135 on a reconnaissance mission in international waters. *Swedish Air Force*

Below:
The Soviet Union's Tupolev Tu-26M, or 'Backfire', bomber, assigned to Russian naval units within operating distance of Sweden, is a familiar sight to the Draken and Viggen pilots of the Swedish Air Force. *Swedish Air Force*

Specification

SAAB AJ37 Viggen
Type: Single-seat all-weather attack aircraft.
Powerplant: One Volvo Flygmotor RM8 two-shaft turbofan engine providing 25,970lb (11,790kg) thrust with afterburing.
Dimensions: Span 34ft 9¼in (10.6m); length 53ft 5½in (16.3m); height 18ft 4½in (5.6m); wing area not disclosed.
Weight: Normal take-off gross weight 35,275lb (16,000kg).
Performance: Maximum speed about 1,320mph (2,135km/hr) at sea level; initial climb rate about 40,000ft (12,200m)/min; service ceiling 60,000ft (18,300m); tactical radius with external ordnance 620 miles (1,000km).
Armament: Seven pylons (option nine) for aggregate external load of 13,200lb (6,000kg) including Rb.04E or Rb.05A air-to-surface missiles and Rb.24, Rb.27 and Rb.28 air-to-air missiles. Capability to carry external 30mm Aden gun pack.

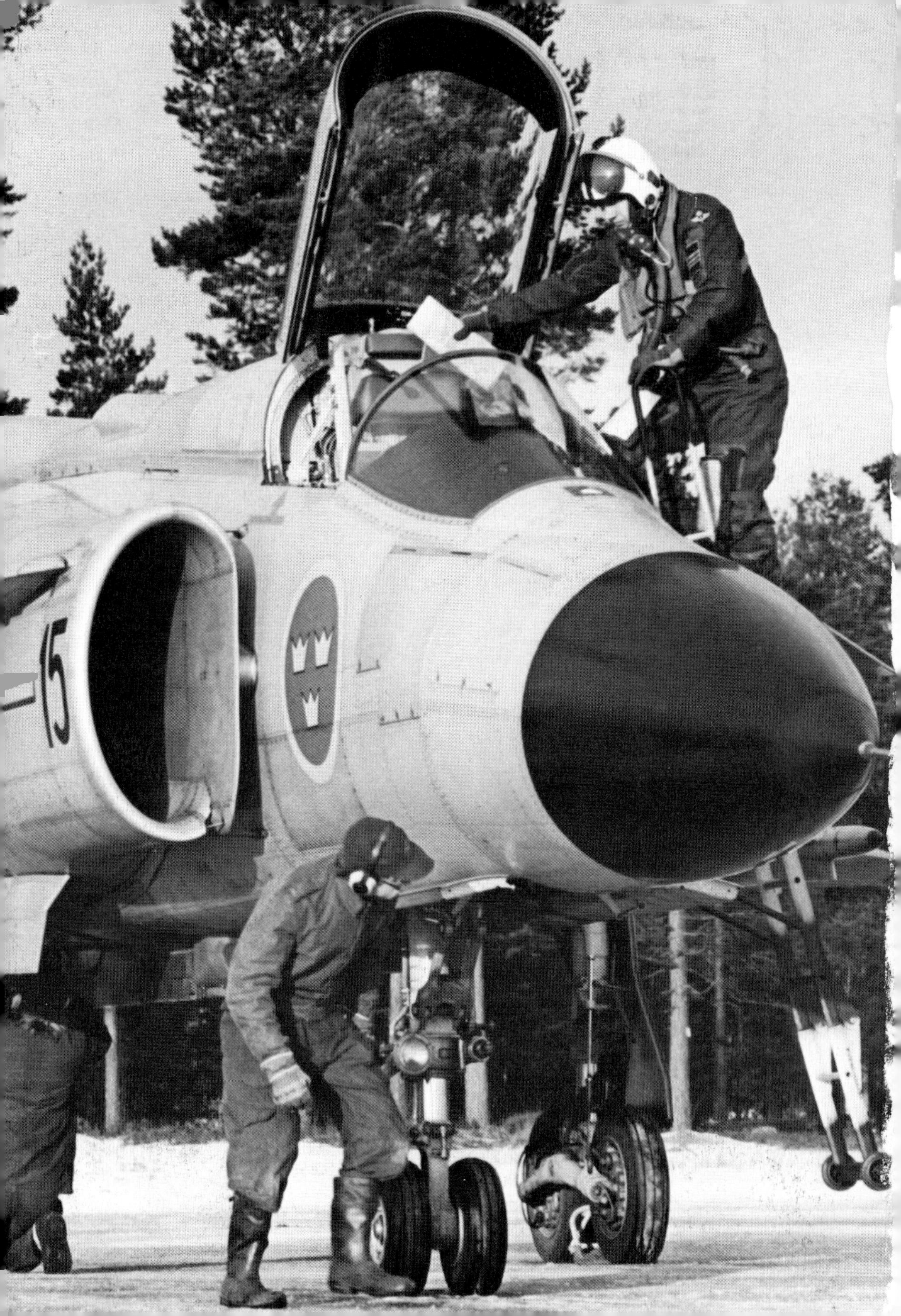
15